Land Pollution Chemistry
An Experimenter's Sourcebook

Hayden Experimenter's Sourcebooks

AIR POLLUTION CHEMISTRY
Herbert Bassow

LAND POLLUTION CHEMISTRY
Herbert Bassow

WATER POLLUTION CHEMISTRY
Herbert Bassow

Land Pollution Chemistry
An Experimenter's Sourcebook

HERBERT BASSOW

Head, Science Department
Germantown Friends School

HAYDEN BOOK COMPANY, INC.
Rochelle Park, New Jersey

Library of Congress Cataloging in Publication Data

Bassow, Herbert.
  Land pollution chemistry.

  Includes bibliographies and index.
  1. Soil pollution.   I. Title.
TD878.B37   333.7   75-19290
ISBN 0-8104-5976-0

Printed in the United States of America

       1   2   3   4   5   6   7   8   9   PRINTING
      ───
         76 77 78 79 80 81 82 83   YEAR

Preface

To say we live in a time of change is both trite and obvious. Our effect upon the planet Earth can be likened to a snowball rolling down a snow-covered hill: slowly at first, then growing in size and speed, becoming more and more noticeable and increasingly dangerous the further it goes.

Human impact on the natural environment has, until recently, been minute. Now, suddenly, we are aware that environmental damage may—if unchecked—destroy us. The damage, like a snowball, grew slowly up to the time of the Industrial Revolution, and with visible rapidity ever since. This visibility has produced awareness, and herein lies our hope. For with awareness can come understanding, and with understanding we can travel the road to meaningful solutions.

It is easy to become cynical and discouraged over the difficulties of initiating seemingly obvious solutions to certain man-made problems. A case in point is the growing realization, since 1973, of the threat to our atmosphere's ozone layer from continued manufacture of the so-called Freons so widely used as refrigerants and as aerosol spray propellents. As these compounds, inert here on Earth, slowly diffuse above the stratospheric ozone, this ozone layer no longer shields them from the Sun's ultraviolet radiation. This radiation breaks them into reactive fragments, such as Cl atoms, which, in turn, destroy the ozone itself. Despite warnings given by reputable scientists since late 1973 that such ozone depletion could seriously threaten life as we know it, the culprit Freon compounds continue to be manufactured in great quantities.

In contrast, consider the Atomic Energy Commission's so-called "peaceful nuclear explosions" program of detonating nuclear devices underground to stimulate the flow of natural gas in "tight" rock formations. It was these gas-stimulation shots—code-named Gas Buggy in New Mexico and Rulison and Rio Blanco in Colorado—that led to the first vigorous public opposition to these underground tests in the United States. The opposition has been so intense that Colorado has amended its constitution to forbid such shots without approval of the electorate, and Congress in the current appropriations act has declared that no money shall be spent on such field testing of nuclear explosives.

Thus public awareness can lead to concrete measures. The purpose of this book is to provide some measure of awareness and understanding of the environmental problems we now face. Such understanding will, I believe, help us to make and support the hard choices that must be made if Earth as we know it is to survive. *Land Pollution Chemistry*, along with its companion volumes, *Air Pollution Chemistry* and *Water Pollution Chemistry*, is designed for beginning chemistry students at the high school and freshman college level, thus encouraging them to use their knowledge of chemistry to gain insight into the nature of many of these environmental problems. It will also be of interest and use to the concerned lay reader.

Philadelphia, Pa.   HERBERT BASSOW

Contents

1

Here We Stand

Just ten years ago, unless you lived in or near a big city, the word *pollution* was probably not in your vocabulary. Today, for anyone living in the United States, or in any of the so-called "developed" countries, that word is not only familiar, it is probably one of the most overworked words around.

The frightening thing for us all is that this is as true today for people who live in the country as it is for city dwellers. As an example, consider *The New York Times* story filed from Whiteface Mountain, whose air used to be normally at the purest level anywhere in the country, but is now reported to be as polluted as the air in some major cities. Whiteface is 120 miles from Albany and 260 miles from New York City, and stands as a striking example of the way in which pollution from our crowded cities has spread to hitherto "pure" areas.

Late 1952 saw reports of Southern California's record five straight weeks of eye-smarting, rubber-cracking, plant-wilting siege of so-called Los Angeles smog. Eighteen years later, a Philadelphia newspaper carried a front-page story about how leaves yellowed and fell unusually early in the autumn smog experienced by the "City of Brotherly Love." Such reports have become commonplace in the past twenty years. The steady increase in their number has finally begun to attract some of the attention this disturbing trend deserves.

The Land

Can you remember a time when you took the garbage out, the garbage truck took it away, you promptly forgot about it? It may still happen this way, but the newspapers no longer let you forget it. Since early 1968, when studies of St. Louis' trash problems were published, the enormity of these problems has become ever more apparent. *Chemical and Engineering News* pointed out that the approximately 5 pounds of solid trash each citizen in St. Louis discards every day adds up to 6,250 tons daily in the St. Louis metropolitan area. It all ends up, we are reminded, either as dumps and landfill dissolved in the Mississippi River after rains soak the dumps, or in the air as gases and particulate matter that rise from incinerators.

New York Times reporter David Bird pointed out recently that the big unsolved problem in New York City was how to get rid of the 24,000 tons of

1

solid waste generated there daily (an amount that is increasing at the rate of 4 percent a year). But one hardly needs to read the newspaper stories to become aware of the problem. The volume of trash waiting for the collector, or just blowing around loose in the streets, provides more eloquent evidence. Add to this the new noncorrosive packaging materials, such as one-way bottles, aluminum cans, and plastic containers, all of which end up as refuse, and the danger that we may bury ourselves in our own trash seems very real indeed.

A study by the National League of Cities and United States Conference of Mayors, reported in the June 10, 1973, *New York Times*, notes that America's cities are smothering in garbage, and almost half of them will run out of places to dump their trash within the next five years.

To make matters worse, more and more of our once beautiful land is being tarnished by man. Consider only two examples, one from the west and one from the east coast. George Kennedy's feature story in the *Philadelphia Inquirer* talks of the dream of California becoming a nightmare. The golden beaches have become oil-splotched, sewage-polluted, and overcrowded. The verdant hills are disappearing under the onslaught of developers and road builders. And that famous golden sun is not only frequently invisible under Los Angeles smog, but is now often "smogged" out over San Francisco Bay and the interior valley.

Bayard Webster, of *The New York Times*, reports the same kind of land erosion and pollution of many parts of the national seashores along our coastlines. From Cape Cod, Massachusetts, to Point Reyes, California, members of national seashore supervisory staffs report overcrowding, littering, and erosion by man and nature as threatening large sections of the 600-odd miles of federally controlled coastal recreation preserves.

We are also becoming painfully aware of the pollution caused by the insecticides and fertilizers man introduced to the land to make it more productive. DDT was released for civilian use in 1945. Just twenty years later, Penn State's Dr. John L. George reported that DDT residues were found in Antarctic penguins, seals, and fish, even though DDT was never used in Antarctica, separated by hundreds of miles of water from areas where DDT might have been used. The questions raised by this paradox have not been fully answered, but they illustrate the global impact of man's tamperings with nature.

A farmer friend who raises steers in Lancaster County, Pennsylvania, reported in 1970 that some of his neighbors who used the insecticide Aldrin two years before to eliminate weeds from their cornfields found enough of the residues of this poisonous substance in the livers of their steers to make the steers inedible. Some dairy farmers are having similar problems with the milk their cows produce.

Summary of Present Problems

Man has been extremely successful in altering his environment to suit his needs and whims: killing animals for food and safety; felling forests for shelter, cropland, and construction materials; mining our Earth for fuels and other minerals; damming rivers for power and water; and producing chemicals to combat pests and disease, and to lessen the toil of farming.

Now, suddenly, we realize that each stage of "progress" has almost always come at the expense of our environment. Oil helps run our cars and heavy industry, but oil spills and offshore leaks kill our wildlife and pollute our beaches. Pesticides improve crop yield, and chemical weed killers eliminate much hand labor, but they pollute our streams, and their residues end up in our foods, and hence in ourselves. Cars give us freedom of movement, but they damage the air and clutter our streets. Medicine has all but eliminated infant mortality, enabled people to live longer, and thereby has contributed to a population growth that, at its present rate, our finite Earth simply cannot continue to support. The development of nuclear energy gives us new power sources and awesome new weapons, but it is debatable whether these weapons increase world stability or make anyone more secure, and it is a certainty that weapons testing and nuclear power plants have polluted underground water supplies, raised river temperatures enough to upset the balance of life in them, and increased the amount of radioactivity in the environment.

Young people today are the first generation, Dr. Barry Commoner points out, to carry radioactive strontium-90 in their bones, DDT in their fat, and asbestos in their lungs. *The New York Times'* Gladwin Hill summarized the problems today's young people must face in his feature article written on the eve of the first "Earth Day." Here are some of his conclusions:

1. Population pressure is already becoming unbearable in many U.S. cities, where city planners' "maximum viable density" of 4,000 residents per square mile has been far exceeded. The consequences range from fiscal crises to a new area of medical studies—urban neuroses. Moreover, 100 million more bodies will have to be accommodated in the United States over the next thirty years. The lack of any firm plans for a host of new communities points to further overcrowding of cities.

2. The most critical result of the mounting population, its concentration in cities, and its push toward an ever higher standard of living is the proliferation of combustion processes—in industry, power plants, automobiles, waste disposal. Something approaching 200 million tons of contaminants, according to federal officials, are now hurled into the atmosphere every year. They run the gamut from black soot of industrial smoke stacks to the colorless, odorless, but potentially lethal carbon monoxide produced by automobiles.

3. Each second of each day and night, about 2 million gallons of sewage and other fluid waste pour into the nation's waterways. No way has been devised to measure the resultant pollution, but nobody has yet contradicted President Johnson's 1967 statement that "every major river system in the country is polluted." The pollution comes mainly from three sources: community sewage, industry, and agriculture. Little has been done about agricultural pollution because it is so diffuse. The abatement of both municipal and industrial sewage has been palpably insufficient.

4. One result of this kind of pollution is that eighty species of wildlife that once lived in the United States are now listed as extinct, and another seventy-eight species—from the timber wolf to the bald eagle—are in danger of extinction.

5. Our inability to cope with our own environmental problems is illustrated, Mr. Hill suggests, by the mounting mass of solid waste around us. Every person in the United States generates an average of 7 pounds of trash a day. Only some 5 pounds of this, according to federal officials, is collected—about 3 pounds of household rubbish, 1 pound of commercial rubbish, 0.6 pound of industrial waste, and 0.2 pound of debris from construction—adding up to about 530,000 tons a day in the U.S. alone!

6. Man's basic heritage—the planet Earth itself, its land areas, the seas around them, and the atmosphere—is being corrupted to an indeterminate degree by a constant infusion of alien chemicals: pesticides, herbicides, fungicides, defoliants, fertilizers, detergent residues, radioactive materials, salts accumulated from irrigation. These substances make their way into lakes, rivers, and the sea, making inroads into certain species, disrupting the natural plant-fish-bird-animal food chain, and perhaps damaging one of the basic processes by which man survives: the conversion by microscopic marine life of waste carbon dioxide into oxygen.

7. Of the 1½ million tons of DDT man has produced, scientists have estimated that two-thirds is still chemically active on the land or in the seas. Some other substances, such as detergent phosphates, are nominally benign; but in waterways they may endlessly stimulate objectionable plant life that tends to transform bodies of water into swamps. This chemical pollution highlights several of the current environmental problems. One is the question of balance between advantages brought by modern science and the ecological price. Another is the question of information: How much needs to be known before scientific advances are exploited? Last is the matter of the decision-making processes to weigh these questions.

This capsule summary of the present state of our environment is not a cheerful picture by any means, but not hopeless either. The first step in attempting to deal with any problem, quite obviously, is to recognize it.

In our attempts to understand the situation today, we shall presently trace, in some detail, the ways in which the present environmental crisis developed. Before attempting this, however, we need to know something about the natural environment on Earth before man appeared and began to change it. It is to this ancient history that we will turn in Chap. 2.

Questions

1. Pick a square block area in a thickly populated residential section of your community, and list every different type of refuse (litter) seen, describing the approximate amount of each kind of litter. This activity should be repeated in various sections of your community.

2. Attempt to learn (from the Sanitation Department) the estimated weights of solid waste collected per day in your community.

Suggested Reading

Garrett De Bell, Ed., *The Environmental Handbook*. Ballantine Books, New York, 1970. A Ballantine-Friends of the Earth Book.

A collection of reprints and some original articles dealing with environmental problems, their causes, and possible solutions. Included are sections on waste disposal, wilderness lands, economics, environmental education and tactics, the population crisis. Bibliography of books, films, environmental organizations.

Wesley Marx, *The Frail Ocean*. Ballantine Books, New York, 1967. A Ballantine-Sierra Club Book.

A carefully written account of some of the ways man has polluted the ocean: beach erosion, effects of sewage and other wastes, including pesticides, use of the ocean by the military. Stress is on understanding the ocean—both its wealth and its tolerances. Detailed bibliography of original sources at end of book.

Robert Rienow, and Leona Train Rienow, *Moment in the Sun*. Ballantine Books, New York, 1967. A Ballantine-Sierra Club Book.

A careful history of how man's ignoring of ecological principles, expanding population, and exploitation of resources has produced the environmental deterioration of land, sea, and air. Notes and chapter-by-chapter bibliography at end of book.

Bibliography

1. David Bird, "Smog Now Found in Rural Regions," *The New York Times*, May 24, 1970.
2. Gary Brooten, "Early Autumn Here Was Result of Smog," *The Evening Bulletin*, Philadelphia, April 16, 1970.
3. Richard Hall, "Air Pollution," *Life*, February 7, 1969.
4. Wilbert C. Lepkowski, "Environment," *Chemical & Engineering News*, March 11, 1968, pp. 9A–17A.
5. David Bird, "Problem of Ridding City of Garbage Eludes a Solution," *The New York Times*, March 24, 1970.
6. George Kennedy, "Pollution Tarnishes Golden State," *Philadelphia Inquirer*, May 14, 1970.
7. Bayard Webster, "National Seashore Parks Struggle With Vast Crowds and Litter," *The New York Times*, May 31, 1970.
8. Research Reporter, "The Pesticide Migration to Antarctica," *Chemistry*, Vol. 38, No. 8, August, 1965, p. 23.
9. David Bird, "Beaches Here Are Found Still Unsafe," *The New York Times*, June 9, 1970.
10. Agis Salpukas, "Mercury Spills Imperil Erie Fisheries," *The New York Times*, May 11, 1970.
11. Science & The Citizen, "Soapless Opera," *Scientific American*, Vol. 189, No. 1, July, 1953, p. 48.
12. Associated Press, "Federal Water Supply Survey Shows 30% of Samples Tainted," *The New York Times*, May 21, 1970.
13. Gladwin Hill, "Man and His Environment," *The New York Times*, April 20, 1970.

2

The Past:
The Pre-Technology Environment

The oceans contain some $1.3 \cdot 10^{14}$ tons of CO_2, about 50 times as much as the air. While some CO_2 is dissolved in the water as CO_2 gas, most is present in carbonate compounds—either as soluble bicarbonates, or as insoluble calcium and magnesium carbonate. The oceans are thought to exchange about 200 billion tons of CO_2 with the atmosphere each year, but this varies. If atmospheric concentrations rise, the oceans tend to absorb much of the excess; when concentrations fall, the oceanic reservoir replenishes them, thus minimizing atmospheric CO_2 fluctuations. Both atmosphere and oceans also exchange CO_2 with rocks and living things (1).* Moreover, CO_2 is gained from volcanic activity that releases it from the Earth's interior, and from respiration and decay of organisms. It is lost to rocks as they weather, and to plants as they use it in photosynthesis (see Fig. 2–1).

Evidence from rock strata studies (2) indicates that for the past billion years, world climate has been largely tropical, interrupted every 250 million years or so by the relatively short glacial periods that bury much of the Earth under gigantic ice sheets. It has been calculated that a 50 percent decrease in atmospheric CO_2 content will lower the average Earth temperature almost $7°F$. Such CO_2 decreases might have been induced in the past by factors such as abnormal uptake by exceedingly flourishing mantles of vegetation. The accompanying temperature drop (a few degrees over a long time would do it) would cause glaciers to spread across the Earth, including the oceans; and since ice can hold little CO_2 as carbonates compared to liquid water, the shrunken oceans would accumulate a CO_2 excess they would eventually have to release to maintain their balance. As CO_2 returned to the atmosphere, Earth's temperatures would rise, and the ice sheets would melt away. This would fill the oceans to their former levels, they would re-absorb the CO_2 they had earlier released, and a new glacial epoch might again begin. Dr. Plass (2) has found strong correlation between scientists' predicted times for such recurrences and the observed times for past glacial cycles.

---

*Literature cited is listed at the end of each chapter.

6

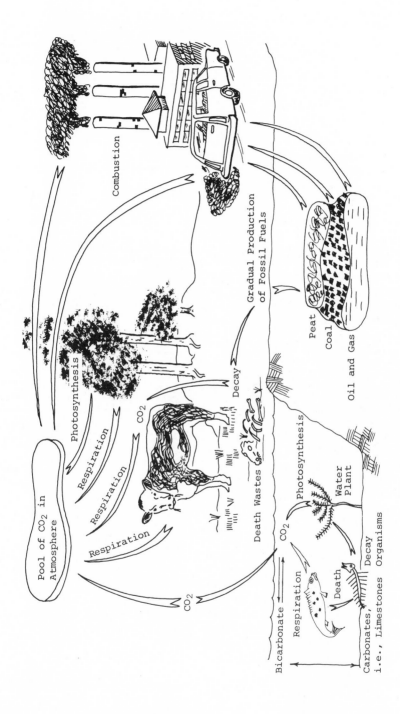

Fig. 2–1. The carbon, or carbon dioxide, cycle. Solid arrows represent flow of CO_2. From Ehrlich and Ehrlich (1).

The Land and Seas: The Carbon and Hydrologic Cycles

In considering the four interconnected carbon dioxide reservoirs (oceans, atmosphere, rocks, and living things), we saw how intimately tied are land, sea, and air. The well-known chemical equilibrium between gaseous and dissolved CO_2 and carbonic acid (H_2CO_3) can be used to describe equilibrium processes occurring in the oceans:

$$\underset{\text{atmospheric}}{CO_2} \rightleftharpoons \underset{\text{ocean dissolved}}{H_2O + CO_2} \rightleftharpoons H_2CO_3$$

But of course the process doesn't stop there because the carbonic acid can undergo acid-base reactions with water to form bicarbonates and carbonates:

$$H_2CO_3 + H_2O \rightleftharpoons H_3O^+ + HCO_3$$

$$HCO_3^- + H_2O \rightleftharpoons H_3O^+ + CO_3$$

Finally, the carbonate precipitates with calcium (and also magnesium) ions in the ocean to form great limestone deposits:

$$Ca^{++} + CO_3^{--} \rightleftharpoons CaCO_3 \text{ (s)}*$$

This calcium carbonate ($CaCO_3$) is the shell material of much marine life. Actually, all these processes are in equilibrium with each other, and together make up the unique carbon dioxide balance.

The familiar demonstration of exhaling through a straw into a small beaker of clear, saturated limewater solution illustrates these processes. As you exhale, the limewater (a solution of calcium and hydroxide ions) becomes milky as insoluble calcium carbonate forms. The steps are represented by the equations written above: the CO_2 dissolves to form H_2CO_3, which reacts with the water to form CO_3^{--}; this in turn reacts with the Ca^{++} in the limewater to form the insoluble $CaCO_3$ precipitate. If you repeat the process, using a fresh sample of limewater, first adding several drops of bromthymol blue indicator, you will notice the initial blue color caused by the hydroxide ions of the limewater changing to yellow, indicating the acidic nature of the resulting solution, as more and more H_2CO_3 is formed.

We also recognize that there exists a vast circulation of water from the sea to the land and back to the sea—the so-called "hydrologic cycle." The total amount of rain and snow falling on the Earth each year is estimated at over $1 \cdot 10^{17}$ (a hundred million billion) gallons! Some 80 percent of it falls over the oceans, with the remaining 20 percent falling over land. A bit more water evaporates from the ocean than falls back into it as rain, but this is balanced by an excess of rainfall over evaporation upon the land. Hence a tremendous volume of water—some $9 \cdot 10^{15}$ (almost ten million billion) gallons—is carried back to the oceans each year by glaciers, rivers, and coastal springs. Of the rain that falls on the land, some is returned directly to the atmosphere by evaporation; some flows off through streams; some is locked more or less permanently

---

*The (s) signifies a solid substance; (g) signifies a (normally) gaseous substance.

in chemical combination with minerals; and some filters through the soils and rock strata of the Earth and eventually returns to the streams or to the ocean.

Groundwater

The amount of fresh water that is stored by all the world's lakes, rivers, streams, and other surface reservoirs is far less, according to geologist A. N. Sayre (3), than that great pool of water stored and transmitted through the porous rocks below the land surface. This pool is called *groundwater*, and the water-bearing formations that make it up are called *aquifers*. Some aquifers (the word is from the Latin, meaning water-bearing) are thousands of square miles in area and hundreds of feet thick; others have a thickness of only a few feet and extend less than a few square miles. The amount and availability of water in an aquifer depends not only on its size, but also on the *porosity* of the rock; i.e., the degree to which it is capable of holding water; and on the *permeability*, or freedom with which water can move through the rock (see Fig. 2–2).

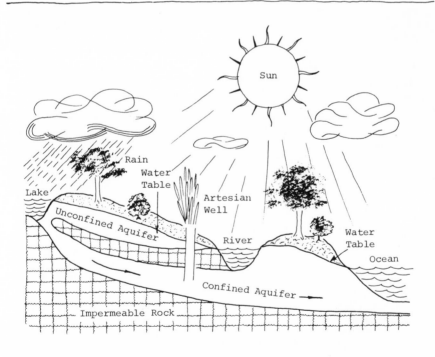

Fig. 2–2. Groundwater and aquifer formations, based on a diagram of A. N. Sayre (3).

According to one estimate (4), the top half-mile of the Earth's crust contains some 3,000 times more fresh water than all the rivers combined. While the true origin of this groundwater is not completely understood, most of it is believed to come from precipitation. Some groundwater may originate in the deeper layers of the Earth's crust by direct combination of hydrogen and oxygen, or perhaps through expulsion of water from the Earth's molten core. When water falls on dry soil as rain or snow, it attaches itself to soil particles by molecular or capillary attraction. The maximum amount the soil can hold is called *field capacity*. When this maximum is reached, runoff occurs; the excess water runs off the land or drains down through spaces between soil particles. Eventually, it fills the hollow spaces in any porous rock layers below, creating a saturated zone when all the hollow spaces have been filled.

The top of this saturated zone is called the *water table*. The water table is almost never a flat surface, but is higher under hills than in valleys. The water table rises and falls as water is added to, or taken from, the zone of saturation. Where the water table intersects the land surface, springs occur or a lake may be formed. Water above the water table can move only downward, but water in the saturated zone below the table moves horizontally until it reaches an outlet. Groundwater, from mountain runoffs, rain, lakes, can only accumulate in aquifers, where a layer of rock contains hollows, or *interstices*, sufficiently interconnected to allow the flow of water (Fig. 2–2). Sand and gravel beds, alluvial deposits, sandstone formations, and porous limestone and basalt all make good, porous aquifer material.

It is important to distinguish between types of aquifers. In an *unconfined* aquifer, the upper limit of saturation, or water table, is the maximum height to which water can rise in a drilled well. Some aquifers, however, are trapped and sandwiched between impermeable rock layers, and are called *confined* aquifers. Because it is pressed upon by the impermeable rock above it, the water in such aquifers is under pressure. When tapped by natural forces, the accumulated pressure from a confined aquifer can drive the water upward as a spring or geyser; or, when drilled by man, can create an *artesian well*.

Under North Africa's Sahara Desert exists a vast aquifer, covering an area of some 200,000 square miles. Recent carbon-14 dating studies have shown (4) that some of this water may have traveled 30,000 years from where it originally entered the ground to reach its present location. It is generally believed that the Sahara was once a tropical region with heavy rainfall, substantial rivers, and abundant vegetation. According to Robert P. Ambroggi (5), most of the water now present in this aquifer beneath the desert was absorbed in past millenniums when the desert received substantial amounts of rainfall.

No discussion of water's role in nature would be complete without considering how much the natural environment has been shaped by the unique properties of water. The heat capacity of water acts as a climate-moderating agent, and the evaporation of *fresh* water alone requires some $6 \cdot 10^{19}$ calories (cal) of heat, which is stored in, and eventually returned to, the atmosphere when precipitation takes place. The hydrologic cycle, however, involves evaporation from the oceans as well. This amounts to some $1 \cdot 10^{17}$ (a hundred million

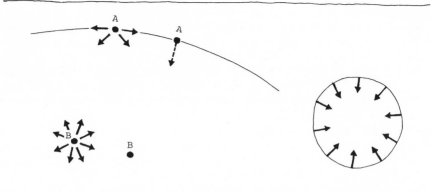

Fig. 2—3. Surface tension in a drop of water.

billion) gallons per year. In all, four times $6 \cdot 10^{19}$ (60 billion billion) is released each year to the atmosphere as a result of precipitation over the entire surface of the Earth.

Raindrops, nearly spherical in shape, have an unusual hardness and strength because of water's high surface tension. All liquids have surface tension, but water's is abnormally high because of the strong hydrogen bonds that join the water molecules together. As shown in Fig. 2—3, the surface molecules (A) in a drop of water are attracted to molecules below and on either side, as shown in the magnified portion at the left. Since there are no attracting molecules above them, they experience a net force directed toward the center of the drop. This net force, experienced by molecules on the surface only, tends to draw water into the compact spherical shape shown at the right. Molecules beneath the surface (B) are attracted equally in all directions by neighboring molecules, which surround them on all sides, and hence feel no net force. The same strong bonds that give water its high surface tension also account for the unusually high boiling point of water [$100°C$, compared to $-61°C$ for its nearest chemical relative, hydrogen sulfide (H_2S)], as well as its high heat of evaporation, nearly 600 cal per gram.

When rain falls on rock and soil, especially when driven by high winds, the hard, tough raindrops make effective bullets which shatter and abrade rock and soil. This abrasive effect is coupled with water's almost unique property of expanding when it solidifies, so that water in rock crevices actually splits granite boulders apart as it freezes. It becomes easy to see the large and dramatic role water plays in changing the face of the land through erosion.

Finally, the remarkable solvent properties of water are believed responsible for dissolving much of the rocks and minerals on the land, and carrying them ultimately to the seas. This, over billions of years, is how the salt, as well as most other mineral deposits, got into the oceans. Seawater contains an average of 35,000 ppm of dissolved solids. In a cubic mile of seawater, weighing 4.7 billion tons, there are therefore about 165 million tons of dissolved matter, mostly chlorine and sodium. The volume of the world's oceans is about 350 million cubic miles, giving a theoretical mineral reserve of about 60 quadrillion tons, quite obviously Earth's largest continuous ore body (see Table 2–1). The abrasive force of flowing water in rivers and streams, aided by already loosened solid materials in dislodging ever more solid rock and soil, is responsible for carrying much insoluble material into the oceans, along with the dissolved solids.

Table 2–1. Concentration (Tons per Cubic Mile) of 57 Elements in Seawater*

Chlorine	89,500,000	Indium	94	Silver	1
Sodium	49,500,000	Zinc	47	Lanthanum	1
Magnesium	6,400,000	Iron	47	Krypton	1
Sulfur	4,200,000	Aluminum	47	Neon	.5
Calcium	1,900,000	Molybdenum	47	Cadmium	.5
Potassium	1,800,000	Selenium	19	Tungsten	.5
Bromine	306,000	Tin	14	Xenon	.5
Carbon	132,000	Copper	14	Germanium	.3
Strontium	38,000	Arsenic	14	Chromium	.2
Boron	23,000	Uranium	14	Thorium	.2
Silicon	14,000	Nickel	9	Scandium	.2
Fluorine	6,100	Vanadium	9	Lead	.1
Argon	2,800	Manganese	9	Mercury	.1
Nitrogen	2,400	Titanium	5	Gallium	.1
Lithium	800	Antimony	2	Bismuth	.1
Rubidium	570	Cobalt	2	Niobium	.05
Phosphorous	330	Cesium	2	Thallium	.05
Iodine	280	Cerium	2	Helium	.03
Barium	140	Yttrium	1	Gold	.02

*From Wenk (6).

Balance in Nature and Biogeochemical Cycles

We have examined in some detail the hydrologic cycle, the complex series of pathways through which water circulates on the Earth. We have also taken a close look at the carbon cycle, perhaps more accurately called the carbon dioxide cycle, for it is the CO_2 in the atmosphere and oceans that constitutes the major reservoir of carbon on Earth. As shown in Fig. 2–1, it is photosynthesis that forms the principal path by which carbon is withdrawn from the CO_2 reservoir and used by plants to form carbohydrates and other organic compounds. The carbon then transfers to the animals who eat the plants and their products, and who replenish the CO_2 again through respiration, using oxygen provided by the plants in the process. Animals also return carbon to the cycle via

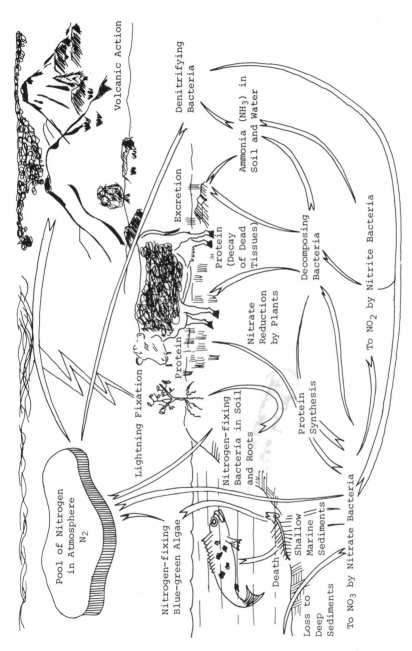

Fig. 2-4. The nitrogen cycle. From Ehrlich and Ehrlich (1).

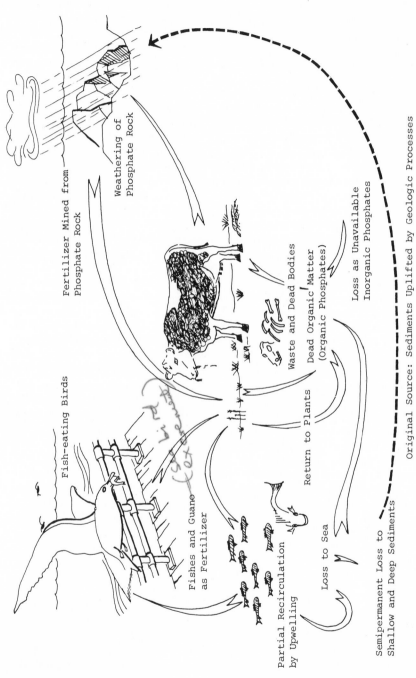

Fig. 2–5. The phosphorus cycle. From Ehrlich and Ehrlich (1).

their droppings, and, after death, through decomposition by microorganisms. Some of the carbon taken in by plants was transformed by geologic processes into fossil fuels—coal, oil, and natural gas—and thus removed from the cycle for millions of years. We have in modern times been returning much of this carbon to the atmospheric pool as CO_2, as we continue to use up those supplies of fossil fuels in combustion processes.

Two highly important biogeochemical cycles not as yet mentioned are those involving nitrogen and phosphorus, both essential to life. The atmosphere is almost 80 percent nitrogen (N_2). While this form is chemically so inert as to be unusable, it does combine into useful compounds by several natural processes. Some microorganisms, for example certain bacteria and blue-green algae, can oxidize the N_2 all the way to nitrate (NO_3^-) ions. Other microorganisms can use atmospheric nitrogen to make their own proteins, which become available to plants when these bacteria die, and to animals when they eat the plants. Bacterial decay of dead plants and animals, as well as animal droppings, produces ammonia (NH_3), which is attacked by a special group of bacteria, and is thus oxidized to nitrite (NO_2^-) ions. The nitrites are then oxidized by still other bacteria to nitrates, the commonest form used by plants. Most fertilizers contain nitrates.

Another group of bacteria reduces nitrates and nitrites, and oxidizes ammonia, to gaseous nitrogen (N_2), which is returned to the atmospheric pool. An important "fixer" of atmospheric nitrogen is lightning. Each year more than 100 million tons of useful nitrogen compounds are spread over the Earth by lightning. Figure 2–4 shows the complex pathways of this so-called nitrogen cycle.

The last of the natural biogeochemical cycles to be examined is the phosphorus cycle. Phosphorus compounds are the important energy-manipulating devices of living cells, and are essential parts of the DNA and RNA molecules that transmit genetic information from generation to generation in all living species. Phosphorus does not cycle as readily as nitrogen or carbon dioxide. Its principal reservoirs are phosphate rocks, deposits of fossilized animals, and *guano* (sea bird excrement) deposits. Although in the same chemical family as nitrogen, phosphorus is so reactive that it is never found free, but as oxide (P_4O_{10}) or phosphate (PO_4) compounds. Phosphates make excellent fertilizers, and are absorbed by plants, and in turn by animals. They return, eventually, to the soil, in animal droppings, and when the plants and animals die and decay. Figure 2–5 illustrates the various routes of phosphorus in its cycle.

Questions

1. Earth's atmosphere is believed to have changed dramatically during its approximately 4-billion-year life. Why does our present atmosphere contain only small amounts of the initially plentiful carbon dioxide?
2. Summarize the process of photosynthesis, emphasizing its cyclic nature.
3. Outline the effects of water vapor and carbon dioxide on Earth temperatures.

4. Write brief summaries of each of the following biogeochemical cycles:
   (a) hydrologic (including ground water)
   (b) carbon
   (c) nitrogen
   (d) phosphorus

Suggested Reading

Harrison Brown, *The Challenge of Man's Future*. Viking Press, New York, 1954.
   While emphasis is on population size and available resources, the first two chapters give an excellent perspective of man's early interaction with his environment.
George Gamow, *Biography of the Earth*. New American Library, New York, 1948. A Mentor paperback.
   Readable account of the birth and evolution of our planet. Chapter 8, "Climates of the Past," gives a good summary of atmospheric changes.
Scientific American, *The Planet Earth*. Simon & Schuster, New York, 1957. A Scientific American Book.
   Reprints of Scientific American magazine articles about our Earth. Parts 4 and 5 include excellent background articles on glaciers, ocean circulation, atmospheric circulation, and the ionosphere.

Literature Cited

1. Paul R. Ehrlich, and Anne H. Ehrlich, *Population, Resources, Environment*, W. H. Freeman & Co., San Francisco, Calif., 1970, pp. 51–65, 157–165.
2. Gilbert N. Plass, "Carbon Dioxide and Climate," *Scientific American*, Vol. 201, No. 1, July, 1959, pp. 41–47.
3. A. N. Sayre, "Ground Water," *Scientific American*, Vol. 183, No. 5, November, 1950, pp. 14–19.
4. Research Reporter, "Ground Water," *Chemistry*, Vol. 39, No. 9, September, 1966, pp. 26–28.
5. Robert P. Ambroggi, "Water Under the Sahara," *Scientific American*, Vol. 214, No. 5, May, 1966, pp. 21–29.
6. Edward Wenk, Jr., "The Physical Resources of the Ocean," *Scientific American*, Vol. 221, No. 3, September, 1969, pp. 167–176.

3

From Past to Present:
A Study of Environmental Changes

Introduction

Chapter 1 sketched a brief but disturbing picture of our present polluted environment. Air pollution serious enough to be a health hazard is no longer confined to large cities, but has spread to hitherto pollution-free rural areas. Garbage and trash threaten the continuing existence of our large cities, while our once naturally beautiful countryside is being destroyed by man and his by-products. An exploding population strains our resources and recreation areas, produces an unbelievable volume of trash, and pollutes the air and water. It is hardly surprising that a 1969–1970 survey revealed that some 30 percent of our public water supplies are contaminated.

In Chap. 2 we looked at the pre-technology environment of the unique, lush, semitropical paradise we call Earth. We saw how in early geologic times the atmosphere made abundant plant life possible by providing a then prodigious amount of carbon dioxide. The plants, in turn, are believed to have provided and replenished through photosynthesis most, if not all, of the present oxygen supply. This interdependence, eventually to include animal life, with their intake of oxygen and release of carbon dioxide for re-use by the plants, emerged as the dominant pattern in nature.

Everything depends, ultimately, on everything else; the more detailed the examination, the more obvious is the interdependence. Even the relatively small amounts of ozone, carbon dioxide, and water vapor make vital contributions to the atmosphere's ability to support life. It is essential, as we begin to trace the ways in which man has harmed this natural environment, that we be aware of the special and delicate nature of this balance. Hence our previous concern with the various biogeochemical cycles with which we concluded Chap. 2.

We are now ready to look at how pollution developed, a story which begins with early man's discovery and subsequent use of fire. It is impossible to pinpoint in time when this occurred, but it may well have been some 650,000 years ago in Africa during a period of volcanic activity (1), when some curious ancestor examined a flaming brand ignited by hot volcanic lava. Man's early use of fire added only insignificantly small amounts of air pollution to that caused

17

by volcanoes and lightning-induced forest fires; with the exception of careless hunters adding to the natural forest fire damage, early man did not contribute measurably to environmental deterioration. As long as man was exclusively a hunter of game and fish, as were other animals, and as long as his numbers remained small, he did not materially affect the environmental balance. He was simply a part of it.

This remained true, we believe, through Paleolithic (Old Stone Age) and Mesolithic (Middle Stone Age) times, a period extending roughly from 2 million up to 12 thousand years ago. It continued even though hunting and fishing communities were being established; tools, pottery, and even bows and arrows had been developed; and dogs had been domesticated to help with the hunting. While the small scale of man's activities, consistent with his small population, did not yet have an impact on the environment, it is probable that the seed for future widespread ecological damage developed in Mesolithic times.

Lincoln Barnett (1) reports that the earliest known example of a "factory" especially built for mass production was discovered by Cambridge archeologist Grahame Clark under a drained field at Star Carr, Yorkshire, England, in 1949—1951. There, where an inland lake existed some 12 thousand years ago, Mesolithic man had apparently built a crude stage, 230 yards square, at the water's edge, and organized on it a crude assembly line for manufacturing tools and weapons. The indications are that he used the water for soaking and softening antlers of deer and moose and for discarding refuse, and that he built fires to extract pitch from wood for use in fastening flint heads to spear shafts. Thus was born the assembly-line technique—and with it man's use of his surroundings for more than immediate survival needs. Perhaps even then it was only his limited numbers that made man's impact on his environment so negligible. According to Paul Ehrlich (2), the population around 8000 B.C., by which time man had spread from Africa to occupy the entire planet, was about five million.

Methods of Dating Past Events

It is appropriate at this point to consider how we can possibly date events that occurred long before recorded history. The methods are mostly chemical and rather straightforward. Those known as *chronometric* are based on independent, time-related natural processes, such as the progressive decay of radioactive isotopes. The most famous of these is radiocarbon dating (3). Its story originated in 1939, when New York University physicist Serge Korff sent neutron counters into the upper atmosphere in balloons. He found that cosmic ray bombardment of the atmosphere produces some 2 neutrons per second for each square centimeter of Earth's surface. Dr. Korff then assumed that these neutrons would react with the most abundant constituent of the atmosphere, nitrogen, in a nuclear reaction involving absorption of the neutron and subsequent emission of a proton (Fig. 3—1), thus changing the nitrogen to carbon. The carbon thus produced—note that it is the heavy carbon-14 isotope—is unstable. Having too many neutrons, it therefore spontaneously undergoes radioactive decay. Although the decay mechanism is somewhat complicated, it may be pictured as a

The formation of carbon-14 by reaction of neutrons (n) from cosmic ray bombardment with atmospheric nitrogen. Structures of N and C nuclei are shown below their respective symbols.

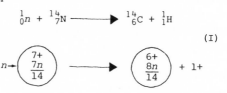

$$_0^1n + {}_7^{14}N \longrightarrow {}_6^{14}C + {}_1^1H$$

(I)

The decay of carbon-14 into nitrogen-14, by emission of a beta-particle. One carbon-14 neutron can be pictured as turning into a proton by emitting an electron (the beta-particle).

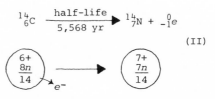

$$_6^{14}C \xrightarrow[\text{5,568 yr}]{\text{half-life}} {}_7^{14}N + {}_{-1}^0e$$

(II)

(A)

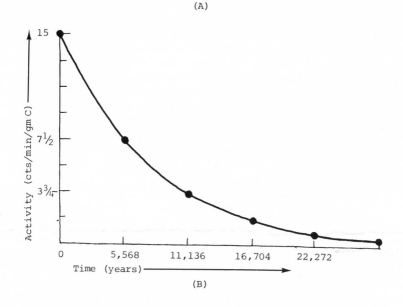

(B)

Fig. 3–1. The chemistry of radiocarbon dating. (A) Nuclear reactions involved in radiocarbon dating. It is assumed that every time a carbon-14 atom is formed (reaction I), an already existing atom of carbon-14 decays (reaction II), thus keeping the amount of carbon-14 and nitrogen-14 constant. (B) Decay curve for carbon-14. Note that its activity (counts/min/gm C) drops to half the former value during one half-life period.

process in which one of the carbon-14 neutrons changes to a proton by emitting an electron (known as a *beta-particle*). Thus, this neutron behaves as if it were a combination of a proton and an electron.

Indeed, when a quantity of free neutrons is created, it has been found that they decay spontaneously, emitting beta-particles (electrons) and becoming protons. It takes some 15 minutes for half of the neutrons to do this. During the next 15 minutes, half of the remaining half (one-fourth of the original number) decay. In the next 15-minute period, half of the remaining one-fourth (one-eighth of the original number of neutrons) decay, and so on. This 15-minute interval is appropriately named the *half-life* of the neutrons. The half-life of carbon-14 (the time it takes for half of the atoms in any sample of it to decay to nitrogen-14) is now generally accepted as 5,568 years.

Willard F. Libby is credited with using this information to develop the technique of radiocarbon dating. He reasoned that this carbon-14 decay has been going on long enough (perhaps ever since the present nitrogen-rich atmosphere was formed some billion years ago) for an equilibrium to have been established. Namely, that for every carbon-14 atom which decays, a new one is created by the cosmic ray-produced neutrons. Thus, the proportion of carbon-14 to ordinary carbon-12 must be constant. Libby then reasoned that that portion of carbon dioxide formed by reaction of carbon with atmospheric oxygen would contain this same proportion of carbon-14 to carbon-12, and hence the carbon-14 would be evenly distributed throughout the atmosphere.

Indeed, because of the equilibrium between atmospheric and ocean-dissolved carbon dioxide, because of photosynthesis during which plants consume carbon dioxide, and finally because animals and men eat the plants (as well as each other), all living things—plant and animal—contain the same portion of carbon-14. This continues as long as the organism lives, because carbon-14 is constantly being replenished from carbon dioxide in the atmosphere. But as soon as an organism dies, or is sealed underground, this exchange stops, and the proportion of carbon-14 to carbon-12 begins to decrease as the carbon-14 disintegrates by beta-particle decay to nitrogen-14.

From a knowledge of the half-life of carbon-14 (5,568 years) and the rate of its formation by cosmic ray bombardment, it has been calculated (3) that the original proportion in living things is one atom of carbon-14 to $0.8 \cdot 10^{12}$ atoms of carbon-12. Such a proportion would produce an average of 15 disintegrations per minute per gram of carbon, or 15 counts per minute per gram of carbon on a Geiger counter. After confirming that this figure is constant all over the Earth (3), Dr. Libby and his associates measured old samples of known age. Wood found sealed up in Egyptian pyramids, dated at about 2700 B.C., for example, should give 7.15 counts per minute per gram of carbon—just about half the activity of a modern wood sample. Actual measurements gave 7.04 counts per minute per gram of carbon, in close agreement with the expected value. By simply recording the counts per minute per gram of carbon in any carbon-containing sample, it thus becomes possible to determine its approximate age. The carbon-14 atoms in the pyramid wood had gone through almost one half-life period (almost 4,700 years). Therefore, its radioactivity should be cut almost in half because about half the original number of carbon-14 atoms are left to decay. Similarly, a carbon-containing object 11,000 years old should retain only one-fourth of its original 15 counts per minute per gram of carbon activity because its carbon-14 atoms have passed through two half-life periods, and thus contain only one-fourth of the original number of radioactive carbon-14 atoms. The actual number of counts per minute differs from these theoretical values because of other factors. As Joan Zimmerman points out, for example, man's burning of fossil fuels (coal, oil, natural gas) over the past hundred years, in which the carbon thus added to the atmosphere as carbon dioxide is old enough to have already lost most of its radioactivity, would tend to dilute and therefore lower the number of counts per minute.

Man Changes the Face of the Land

The Neolithic Age was the crucial turning point in man's history. It was a triumph because he discovered the arts of agriculture and animal husbandry, and thereby acquired control of his environment. But it also marked the beginning of ecological disaster. In shaping his environment man began drastically to change and upset it. Once man realized that he could grow and store more food than he could acquire by hunting, he abandoned the roving, predatory life of the hunter for the relative ease, abundance, and security of farming. Among other results were lowered infant mortality and longer life spans. Human populations began to increase out of all proportion to other mammalian species.

The earliest evidence of plant cultivation yet found is in the highlands and plateaus above the fertile river valleys of Iraq and dates back to 8000 B.C. The first crops were probably wild barley and wheat, with cultivated forms evolving somewhat later. A cultivated form of emmer, which resembles modern wheat, has been dated at about 6000 B.C. at the ancient town of Jarmo in Iraq. According to Lincoln Barnett (1), the original landscape in this area was open and fertile like a park, dotted with intermittent trees and filled with meadows of wild grain. Today this area is desert, completely devoid of its original fertility. By 4000 B.C. this Neolithic farming culture had spread to Europe, moving up the Balkan peninsula to the Danube valley and then westward. There are indications that farming was practiced in Spain by 3500 B.C. and in England and the New World by 2500 B.C.

The discovery of agriculture wrought such great changes that the period 3500 to 2500 B.C., beginning less than 5,000 years after the first plant cultivation, saw the flourishing of the world's first great recorded civilizations. During this period the Sumerians came of age around the Tigris and Euphrates rivers near Iraq, and the Egyptian civilization flowered along the Nile. By 3500 B.C., and during the Bronze age beginning about 500 years later, metals were first mined, copper and later an alloy of copper and tin first being used to make tools and weapons. With the knowledge that heating native ores with charcoal would yield valuable metals, the search for and extraction of mineral resources was on.

Gold deposits were widely distributed in Egypt and other parts of the ancient world. Silver and lead both occur in the then widely distributed galena ores. First the ore (principally PbS) was burned:

Mining pollute. $SO_2 + CO$

$$2PbS + 3O_2 \longrightarrow 2PbO + 2SO_2$$

At higher temperatures, the PbO could be reduced by charcoal (C):

$$2PbO + C \longrightarrow 2Pb + CO_2 \text{ or } CO \text{ (more likely)}$$

The SO_2 and CO may have been among the earliest man-made air pollutants! Native copper supplies in the Near East were quickly exhausted, and man began to extract copper from its carbonate and oxide ores by heating with charcoal:

$$CuCO_3 + C \longrightarrow Cu + CO_2 + CO$$

$$CuO + C \longrightarrow Cu + CO$$

More man-made air pollution! The extraction of tin and iron soon followed. As ore supplies in the Near East became exhausted, the search for new supplies took traders into the Danube Valley and subsequently into middle Europe.

Man's denuding of his environment, and the search for new mineral resources and fertile soil forced by this denuding, was (and remains) the determining factor in his history. The Babylonians and Assyrians accumulated riches from their mineral resources and the fertility of their lands, watered by the great Tigris and Euphrates rivers. While we do not know exactly why these and other great early civilizations in this region declined, we do know that their peoples quickly exhausted their mineral resources and the productivity of their soil. Thus, much of what was once the fertile cradle of civilization in the Near East rather quickly became, and is to this day, an arid desert. The glorious Greek civilization that developed centuries later was also short-lived. Plato, who lived to see its decline, wrote of Attica, once rich in soil and heavily forested mountains, quickly stripped of its forests to provide timber as well as additional pasture for cattle. He complained bitterly about the ravaged land, no longer able to hold or store rainfall, but instead washed into the sea (4). This combination of deforestation, which robbed Greece of its water supply and hence its stable agriculture, and the exhaustion of silver and other minerals from the mines at Laurium, may well have caused the decline of Athens (4). To this day, the land has never recovered.

The pattern continued with Rome (4). As early as the first century A.D., Columella wrote of once-fertile Latium as a place where the population would die but for imported grain. By his time arable lands in Italy had declined to the point where grain farming had to be abandoned, and Rome was forced to turn to her colonies for food supplies. Although the Romans had a knowledge of agricultural practices equivalent to that of nineteenth century Europe, they were apparently unable to maintain sustained and widespread application of these practices. Thus much of southern Italy, once fertile enough to attract colonizing Greeks, became and has remained desolate land. The Romans then conquered Carthage, where they proceeded to ravage the land, mine dry the mineral resources, and cut down the Pyrenees and the Sierra Morena forests, using the wood to build ships and houses.

The factors which cause soil erosion are obvious enough to anyone concerned enough to observe them at work. The Romans certainly were acquainted with them. Most important is the amount of cover provided by trees and grass. When these are removed by cutting and overgrazing, the remaining bare soil is extremely vulnerable to erosion. Each time a driving rain hits such land, bare soil is torn loose and washed away. Alternate freezing and thawing, or wetting and drying, loosens surface soil and breaks down soil aggregates to granules that can easily be swept away by the wind. Heavy cropping and tillage cause organic matter to decompose rapidly, also creating breakdown. The damage occurs in stages: First comes *sheet erosion* loss, where rainwater runoff removes thin layers of surface soil, a layer at a time. Then, the remaining surface soil becomes thinner, small gullies appear, which are then deepened by water drainage, hastening soil loss. Finally, the wind carries away the finer soil particles, which contain the major part of the plant food elements. Only the coarser, less fertile, more erodable soil is left behind.

Donald E. Carr (5) questions why Mesopotamia and the "Fertile Crescent" ended up as desert, despite the knowledge and enterprise of such early peoples as the Babylonians and Assyrians. They understood and used irrigation, built mountain reservoirs, made even once-desert areas bloom luxuriously. Yet these man-made gardens, along with the naturally fertile areas, ended up as a "broken-down dust heap" (6). Historian Arnold Toynbee contends that the Mongol invasion in the fourteenth century A.D. so crippled the Mesopotamians that they were unable to continue their irrigation and reclamation work. Indications are, however, that the land lost its fertility long before any Mongol invasion. Mr. Carr is more inclined to accept the hypothesis of Paul B. Sears, who attributes the decline to the gradual silting up of the entire water system. He feels that the normal silting of the Tigris and Euphrates Rivers was disastrously increased by the practices of overgrazing and overcutting of highland vegetation, whose runoff supplied the rivers. The imminent peril of eroded soil thus silting the river beds may not have been fully appreciated by the Mesopotamians. But their sheepherders, lumbermen, and farmers—along with the rest of the population— felt pressured for immediate results from the land rather than careful long-term planning, a pressure that has been man's curse through the ages. The amount of labor needed to clean the silt from river and canal beds would have left little available for other projects. Thus, in all likelihood, the tired civilization simply gave up on their future.

It is easy to see why man has allowed much of his land to erode away. At first land is plentiful, while labor is not. Immediate needs have always led man to take from the land with little or no thought about long-term effects. Forests are cut to provide wood for houses, or simply to make more land available for crops or grazing. The land is used over and over, until its fertility and cover are gone. The rain erodes it away to silt up rivers or be lost to the sea long before new grasses and trees can sink firm roots to hold it in place. Man then moves to virgin land and proceeds to subject it to the same treatment. We can therefore see what became of the once-fertile Near East, of Greece, of Rome. Old Chinese records tell a similar story. Lured by the profits from wool, Spain, once fertile, allowed the land to fall under the domination of the Mesta, the National Association of Sheepowners, in the fifteenth and sixteenth centuries. For a hundred years, the land was overgrazed and forests were burned off to give extra pasturage; the cycle of overuse and erosion was thus set in full motion. The power of the Mesta lasted long enough to damage severely the agricultural regions of Spain and with it Spain's dominant place in the world. Wherever man has gone on Earth, the story has been the same. It has been repeated in much of North and South America, and Asia. In the United States, much of our once-fertile midwestern Great Plains was in less than 100 years transformed into today's "dust bowl" through overgrazing and uncontrolled cultivation. Basutoland, just north of the Cape of Good Hope in South Africa, was described as a paradise—a fat and lovely land—by nineteenth century French missionaries. Today Lesotho (the name selected by newly independent citizens of Basutoland in 1966) is, in *The New York Times'* reporter Charles Mohr's words (6), "arid, eroded, almost treeless and ravaged." It has been, and still is, overgrazed and overstocked by the cattle the farmers still regard as visible symbols of status, and are therefore reluctant to sell or butcher. "It's the perfect example of what you can do to an ecology in less than a hundred years," observed one resident (6).

Questions

1. Outline the method of radiocarbon dating. If the activity of a wooden spear, found recently in a sealed cave, gives 3.7 counts per minute, what is the spear's approximate age?
2. Consider how man's discovery of agriculture changed the face of the land; how a fertile valley could become a barren desert. List the steps by which such change occurred. Then project alternate methods that would have slowed or eliminated these changes. How may such changes have affected man's history?

Literature Cited

1. Lincoln Barnett, "The Epic of Man," *Life*, Part I, November 7, 1955; Part III, February 27, 1956; Part IV, April 16, 1956.
2. Paul R. Ehrlich, and Anne H. Ehrlich, *Population, Resources, Environment*, W. H. Freeman & Co., San Francisco, Calif., 1970, p. 9.
3. Joan Zimmerman, "Answering the Question When?" *Chemistry*, Vol. 43, No. 7, July–August, 1970, pp. 22–27.
4. Fairfield Osborn, *The Limits of the Earth*, Little, Brown & Co., Boston, Mass., 1953, pp. 7–22.
5. Donald E. Carr, *Death of the Sweet Waters*, W.W. Norton & Co., New York, 1966, pp. 15–28.
6. Charles Mohr, "Lesotho, Once Lush Region, Is Now Parched and Eroded," *The New York Times*, May 21, 1970.

4

State of the Land:
Agriculture and Radioactive Pollution

Introduction

Before man appeared, competition between the many species for places in the natural environment produced a balance, thereby preventing dominance by any single life form. Man's clearing and planting of the land, year after year, often for a single crop, has upset this balance. The concentration of wheat and cotton plants in a given area, for example, facilitates the spread of wheat rust and cotton boll weevil pests. Indeed, whenever cleared land is used for such a specific purpose, it encourages the massive increase of those few species that find the newly altered environment most suitable. The artificial, less complex man-made environment not only produces food crops efficiently—it also produces food crop pests efficiently.

The sad fact is that man's alterations of nature have all too often created imbalances which favor his natural enemies. Millions of acres, cleared and planted, have created a haven and abundant food supply for insect pests and plant disease organisms. The crowding of hundreds of domestic animals into areas where perhaps only dozens would occur in a natural state gives animal parasites an enormous advantage. Worse still is the imbalance created when man unwittingly carries pests into new surroundings. The Japanese beetle, for example, was accidentally introduced into the United States, probably as larvae in the soil around imported plants. First discovered here near Riverton, New Jersey, in 1916, it had soon spread to most Atlantic seaboard states, extending as far west as Ohio. Over half the losses caused by insects in the United States (and hence more than half the insecticides we use) are traceable to insects that were accidentally introduced into this country. With these chance importations, the natural control agents (those organisms that normally feed on these insects) are usually left behind, so that the destructive pests are left without natural checks and therefore, develop huge populations, which cause far greater damage than would be possible in their native homes.

Humans have also reproduced unchecked, as they have increasingly controlled and tamed the once-hostile environment. Having grown from some 500 million people 300 years ago to over 3 billion today, the human population will reach 6 billion by the year 2000, if current rates of population increase remain unchanged (see Fig. 4–1). We have never adequately fed the world's entire human population, and the task of feeding the ever-increasing multitudes is the overriding problem for us all. At present, some 7 percent of the Earth's land area is used for crop production during any given year. Since most unused land is too dry, or too cold, or in some other way unsuitable for agriculture, we might look to tropical and other forest areas as potential additional cropland. The experts, however, tell us that neither extensive clearing of forests nor large-scale cultivation of tropical regions holds much promise. One reason is that imbalance and erosion damage inevitably follow forest denuding. Furthermore, much of the soil

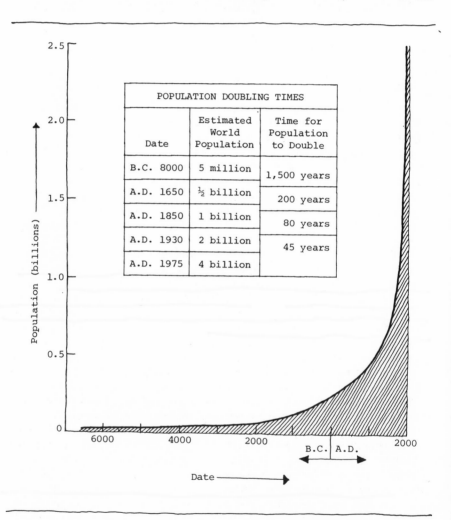

POPULATION DOUBLING TIMES		
Date	Estimated World Population	Time for Population to Double
B.C. 8000	5 million	1,500 years
A.D. 1650	½ billion	200 years
A.D. 1850	1 billion	80 years
A.D. 1930	2 billion	45 years
A.D. 1975	4 billion	

Fig. 4–1. Population growth. Table from Ehrlich and Ehrlich (1).

in tropical regions is *lateritic* and turns hard as the result of an oxidizing effect when put to the plow.

The great Sahara Desert of North Africa is largely manmade, the result of overgrazing, deforestation, faulty irrigation, and weather changes probably influenced by the denuding. Even today the Sahara is reported to be expanding southward by several miles each year. Western India's Thar Desert, jungle just 2,000 years ago, was also created by deforestation and overgrazing, and is also increasing in size each year. It is shocking to realize that the amount of land classified as either desert or wasteland was only 9.4 percent of total land area in 1882, and had risen to 23.3 percent by 1952—largely because of man's activities. The obvious answer, besides trying to limit population growth, is to attempt to increase the productivity of tillable soil presently used as cropland.

Man quickly learned, even in Neolithic times, about the effects of organic (or natural) fertilizers on growing plants, probably from observing the lush growth induced by animal droppings and carcasses. The Romans by about 200 B.C. knew enough about soil fertility to recommend crop rotation, liming of acid soils, manuring, and growing legumes such as clover, which we now know fixes atmospheric nitrogen into the soil. Early English settlers in North America found that the Indians increased maize yields by burying a fish with each seed, while farmers in medieval Europe often grew clover and rotated crops because they had learned from experience that such practices helped maintain soil fertility. By the early 1800s, when bones and blood had been added to animal wastes, it seemed clear that the limited supplies of such organic fertilizers would be insufficient to meet the rising demands for food of an already rapidly expanding population.

Chemical Fertilizers

It was a German chemist, J.R. Glauber, who first demonstrated in the early 1600s that an inorganic chemical, saltpeter or potassium nitrate (KNO_3), produced large crop responses. The Swiss chemist Nicolas de Saussure was able to show in 1804 that plants grow luxuriantly on carbon and oxygen from the air, and mineral nutrients from the soil. By the late 1800s, scientific research, continuing at an accelerated rate, had determined most of the essential elements in plant nutrition, and pointed the way toward replacement of organic fertilizers with chemical substitutes. By the time nitrogen fixation came to be understood in the early 1900s, the rapidly expanding chemical industry was able to provide industrial replacements for organic nitrogen in fertilizer. The manufacture of coal gas by heating coal gave, per ton of coal used, about 5 pounds of ammonia (NH_3) as a by-product, which could then be made into ammonium sulfate [$(NH_4)_2SO_4$]. Mines in Chile began large-scale production and export of sodium nitrate ($NaNO_3$). This helped reduce the proportion of organic nitrogen fertilizers used in the United States from 91 percent in 1900 to 40 percent by 1913. By 1915, the Haber Process for synthesizing ammonia had been perfected, making it possible to produce in one commercial plant 20 tons of ammonia per day, and making ammonia, together with such derivatives as ammonium nitrate (NH_4NO_3) and urea (NH_2CONH_2) available on a large scale.

Although the first association of phosphorus with bones was made back in 1769, it was not until after 1840 that chemists recognized phosphorus as the key ingredient that made animal bones such effective fertilizers. At about the same time, the Great German chemist von Liebig postulated the use of sulfuric acid (H_2SO_4) to make the phosphorus in bones more readily available to plants. By 1865 England alone, acutely in need of additional sources of fertilizers, was producing 200,000 tons a year of phosphate fertilizers, made from sulfuric acid, phosphate fossil manures, and minerals (2). In the 1890s, the slag removed while producing iron and steel from high phosphate ores became yet another source of phosphorus for agriculture. The identification of potassium as the beneficial element of potash, the crude potassium carbonate contained in wood ashes, was made soon after the isolation and discovery of potassium by Sir Humphry Davy in 1807. Potassium fertilizers were in commercial production by the 1860s.

Thus, by the mid-1800s man had identified nitrogen, phosphorus, and potassium as the three so-called "macronutrients" needed by plants in large amounts, and had learned to mass-produce chemical fertilizers to supply them (see Fig. 4–2). Nitrogen accelerates growth, increases yields, and promotes activity of soil bacteria. Phosphorus stimulates germination of seedlings and encourages early root formation. Potassium improves yield, enhances formation of starches, sugars, and plant oils, and increases plant vigor and resistance to frost and disease. Any study of crop yield versus the amount of fertilizer used (Fig. 4–3) makes a compelling argument for fertilizers, even though crop yields level off following repeated applications. For our purposes, however, two things must also be born in mind.

First, these fertilizers are made by industrial processes, which pollute the environment. Phosphate fertilizers, for example, are made largely from phosphate rock, or apatite, which is essentially calcium fluorophosphate [$CaF_2 \cdot 3Ca_3(PO_4)_2$]. The most effective fertilizers result from treatment of this ore with sulfuric acid to produce superphosphate, which is principally $CaH_4(PO)_4)_2 \cdot H_2O$, sometimes written $Ca(H_2PO_4)_2 \cdot H_2O$, and represented by the equation:

$$CaF_2 \cdot 3Ca_3(PO_4)_2 \cdot H_2O + 7H_2SO_4 + 17H_2O \longrightarrow$$

$$3CaH_4(PO_4)_2 \cdot H_2O + 7CaSO_4 \cdot 2H_2O + 2HF$$

The hydrogen fluoride (HF), a deadly poison, is supposedly reacted with SiO_2, but in fact it and other gaseous compounds of fluorine escape into the atmosphere during processing. In Polk County, Florida, to cite one of the many cases, fluoride emissions from phosphate fertilizer plants have wiped out tens of thousands of cattle. A survey of 777 cattle "survivors" showed 71 percent with such fluorosis symptoms as spongy teeth, thickened bones, stiffening joints (3).

Second, the fertilizers themselves can pollute. Nitrate nitrogen fertilizers such as ammonium, calcium, potassium, and sodium nitrates, all widely used throughout the world, yield nitrogen that is quickly available to root systems. Often, therefore, plants grow too rapidly. More importantly, nitrates are so soluble that such fertilizers are easily *leached* (dissolved) out of the soil, entering and polluting groundwater supplies. Indeed, more and more cases of this nature

SOURCES	PRODUCT
Phosphate Rock	Ground Phosphate Rock (0-35-0)
Phosphate Rock Sulfuric Acid	Single Superphosphate (0-20-0)
Phosphate Rock Phosphoric Acid	Triple Superphosphate (0-48-0)
Hydrocarbons Steam Air	Ammonia (82-0-0)
Ammonia Sulfuric Acid	Ammonium Sulfate (21-0-0)
Ammonia Phosphoric Acid	Ammonium Phosphate (18-46-0)
Phosphate Rock Nitric Acid Ammonia	Nitrophosphates (20-20-0) Calcium Nitrate (15-0-0)
Ammonia Nitric Acid	Ammonium Nitrate (33-0-0)
Ammonia Carbon Dioxide	Urea (46-0-0)
Ammonia Carbon Dioxide Salt	Sodium Carbonate Ammonium Chloride (23-0-0)
Coal Limestone Nitrogen	Calcium Cyanamide (24-0-0)

Fig. 4–2. Basic components used in manufacturing the major kinds of chemical fertilizers. Numbers in parentheses show, respectively, the typical percentages of nitrogen, phosphorus, and potassium materials in each fertilizer. After Pratt (2).

are now coming to light. George Kennedy cites two cities in California's San Joaquin Valley whose water supplies have been certified as unsafe for drinking by infants because nitrate levels in the underground water supplies are twice the amounts considered unsafe by State and Federal authorities (4).

How did the water get this way? From the use of chemical fertilizers to increase crop yields, for these nitrates are residues of nitrogen fertilizers used by

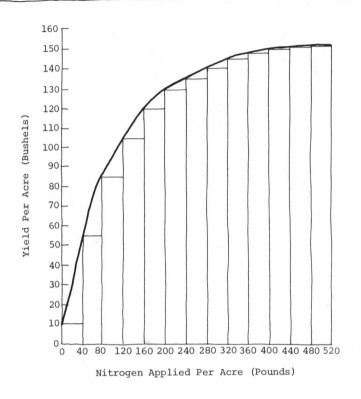

Fig. 4—3. Economics of fertilizer use.

area farmers for generations. Through constant irrigation, the water table has risen to the level of the nitrate nitrogen deposited in the soil, dissolved it, and carried it into the wells. One of the important sources of pollution in unbelievably filthy Lake Erie is the runoff from the estimated 30,000 square miles of farmland in the Lake Erie Basin. The waters draining from these lands are rich in nitrogen from the repeated heavy use of soluble nitrate fertilizers. It has been estimated that the nitrogen content of these waters from fertilizer runoff is roughly equivalent to the sewage of some 20 million people—or about twice the population of the Lake Erie Basin!

Mineral Insecticides and DDT

We have already mentioned that increased food crop production also means increased crop pest populations. The Colorado potato beetle, once an insignificant insect feeding on wild herbs in the Rocky Mountain region, has grown abundant as a result of the cultivated potato crop. The alfalfa caterpillar, first discovered in California about 1850, lived on native plants such as clover and locoweed. With the subsequent development of alfalfa farming in California, now involving over a million and a half acres, the caterpillar population has mushroomed. A single female butterfly, the adult stage of this caterpillar, can lay as many as 1,500 eggs; unchecked, this *one* insect would soon bury us!

Insects, according to latest estimates, comprise some 3 million species, which is more than all other plant and animal species combined! The number of individual insects alive at any one moment is estimated to be a billion billion, or 10^{18} insects. Written out, this becomes 1,000,000,000,000,000,000! Of this vast multitude, 99.9 percent are either harmless or beneficial to humans, while the other 0.1 percent, amounting to some 3,000 species (or 10^{15} individual insects), are the agricultural pests and disease carriers. Since these would indeed wipe out our food crops if left uncontrolled, and since man has eliminated their natural controls by altering the environment, it is obviously necessary for man to impose other methods of control.

The earliest insecticides, first used widely in the eighteenth century, were of vegetable origin, such as the nicotine preparations made by steeping tobacco leaves in water. Mineral insecticides, chiefly poisonous compounds of copper and arsenic, were introduced in the nineteenth century, when chemicals such as arsenate of lead [$Pb_3(AsO_4)_2$] came into general use. Because these compounds are known poisons, they have been used with some care.

The majority of synthetic insecticides in wide use since the 1940s are either chlorinated hydrocarbons or organophosphates. The most widely known, of course, is DDT. Interestingly, DDT, whose chemical name is *di*chloro-*di*phenyl*tri*chloroethane (Fig. 4–4), first synthesized by German Ph.D. candidate Othmar Zeidlar as an exercise in pure chemistry, remained unrecognized as an insecticide for 65 years. Its lethal effect on insects was detected in 1939 by the Swiss chemist Paul Mueller, who was searching for a chemical that could kill insects on contact. His work with DDT brought him a Nobel prize in 1948. The United States Army began testing DDT in 1942, and used it successfully in malaria control and other programs, in which it is estimated to have saved some ten million lives, and to have eliminated perhaps 200 million human illnesses. It was cheap (a dollar per pound, initially, with a present cost of about twenty cents) and persistent; a single spraying often produced lasting effects. Soon after its initial successes with the Army, DDT became available for civilian use, and was put to work in Michigan, controlling flies that threatened fruit farms bordering the Great Lakes, and eliminating pesky mosquitoes. Even Michigan conservationists, unaware of its potential danger, sprayed it in state parks, often located near water. Homeowners and townships sprayed DDT in back yards and streets, from which it drifted into sewers and rivers, and finally into the Great Lakes.

After the recent and continuing barrage of DDT warnings, it is difficult to realize the enthusiasm with which it and other chlorinated hydrocarbons were initially received and used. Not only did DDT prove extraordinarily effective against a wide variety of insects, but it also killed on contact, provided long-lasting protection, and was easy and inexpensive to mass-produce. Mass applica-

Dichlorodiphenyltrichloroethane (DDT)

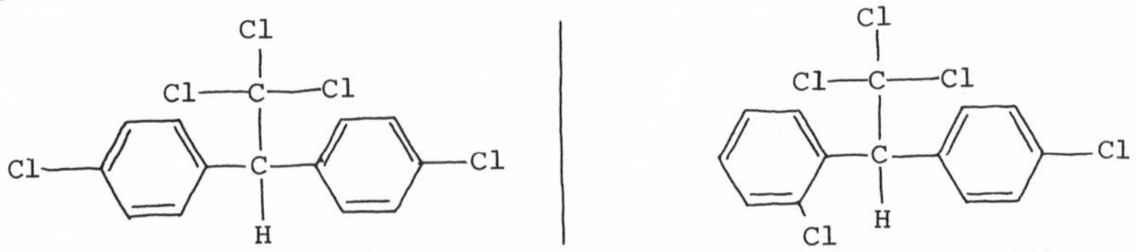

80% *para, para*-DDT   and   20% *ortho, para*-DDT

(A) The chemical "formula" of commercial DDT. Recall two
    groups attached to opposite carbons in benzene ring make
    it "*para-*" (left), while groups attached to adjacent
    carbons result in "*ortho-*" nomenclature (right).
    Commercial DDT is usually 80% *para, para-* and 15 to 20%
    ortho, para-DDT, together with other chemical impurities
    not shown.

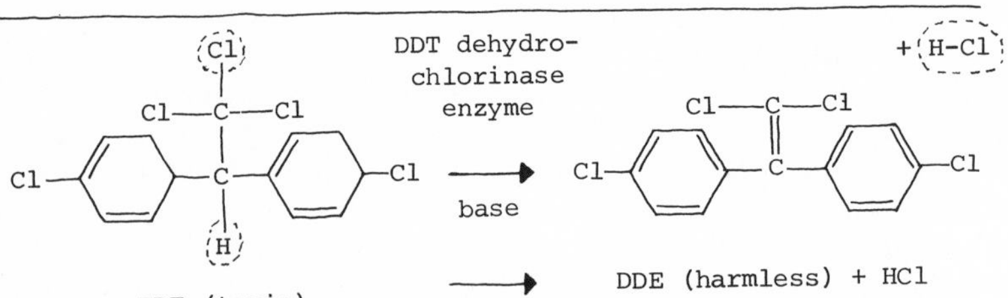

DDT (toxic)   DDE (harmless) + HCl

(B) One way DDT-resistant insects detoxify the insecticide:
    basic enzymes within the insect remove a Cl and H,
    circled on left, which form an HCl molecule, leaving the
    double-bonded ethylene derivative, DDE (right), which is
    relatively harmless to insects, but harmful to birds.

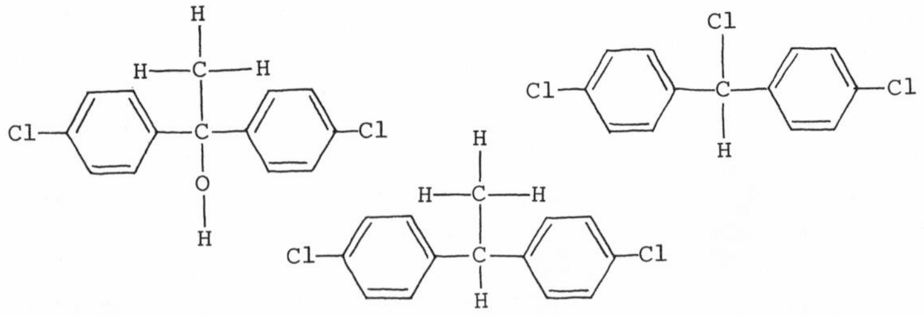

(C) Three substances which have been found to activate DDT
    against resistant flies when mixed with it. These sub-
    stances are harmless when used alone. Note the similarity
    of each to *para, para*-DDT (top left), the major constitu-
    ent of commercial DDT.

Fig. 4—4. Chemistry of DDT.

tions of DDT powder arrested an epidemic of louse-borne typhus in Naples during the initial period of Allied occupation in 1944. After World War II, DDT sprayings virtually eliminated malaria from Sardinia. Wholesale sprayings of towns and villages with DDT practically wiped out flies in the late 1940s. In March 1951, Iran asked the United States to help control swarms of locusts massing along the Persian Gulf, which threatened to destroy that country's entire food crop. The U.S. Army sprayed the area from the air with 10 tons of the chlorinated hydrocarbon Aldrin (Fig. 4–5), and almost overnight wiped out the locusts. For the first time ever, a country-wide locust threat—synonymous since Biblical times with disaster—was nipped in the bud.

Normal losses inflicted on United States crops by pests, weeds, etc. amounts to almost half of the total production. In 1950, for example, crop damage was an estimated $13 billion out of a gross farm production of $31 billion (5). Thus it was hardly surprising in 1951 when widespread concern greeted the World Health Organization's March report that a developing shortage of DDT was serious enough to threaten campaigns against insect-borne disease. At that time DDT was viewed as one of the world's essential needs. United States production of DDT in the 1950s, some 65 million pounds yearly, then represented about 75 percent of the world's supply. Commenting on the 1951 shortage, John R. Murdock, assistant director of the World Health Organization's Pan American Sanitary Bureau, warned that unless production was increased, all gains in control of malaria, typhus, and plague since 1947, affecting some 300 million people in tropical countries of the world, would be wiped out. El Salvador's Under-Secretary of Public Health, Roberto Caceres Bustamente, saw the DDT shortage as a life-or-death problem for his 2.5 million people. Although El Salvador still had over 200,000 cases of malaria, Bustamente said that DDT had reduced its incidence by 40 percent (6). Even today, the World Health Organization maintains that no adequate substitute for DDT exists in controlling malaria in developing countries, and feels that its immediate discontinuation would be a disaster. By mid-1970, the organization had reportedly screened 1,300 possible substitutes for DDT in malaria control, and found only two with promise—both of them costlier and less persistent than DDT (7).

Along with DDT's life-saving benefits, previews of coming problems were emerging. Biologist Barry Commoner, tells of spraying a New Jersey coastal island with DDT, while still serving in the U.S. Navy during World War II, to protect an experimental rocket station from flies so numerous that they actually delayed work. Within hours the flies were dead, but a week later tons of decaying fish littered the beach, apparently also killed by the DDT. This in turn attracted swarms of new flies from the mainland (8). Another example cited by Commoner occurred after the war in a Bolivian town sprayed with DDT to control malarial mosquitoes. It killed the mosquitoes, but most of the local cats as well. With the cats gone, black typhus-carrying mice invaded the town, and before new cats could be imported to restore the balance, several hundred villagers died from typhus.

As early as 1948, housewives, dairymen, and hog farmers throughout the United States began to notice that DDT no longer killed houseflies and mosquitoes with its original potency, and blamed manufacturers for skimping on their

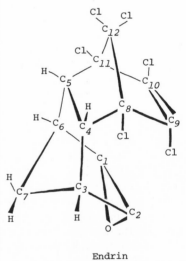

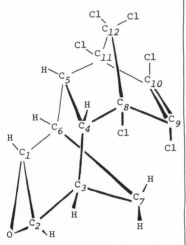

Dieldrin

Endrin

product. The blame, however, lay with the insects, who developed surprising immunity to DDT through fundamental biochemical and genetic processes. Even in 1947, reports from Switzerland and Italy indicated strains of insects, hundreds of times more resistant to DDT than their ancestors, living and reproducing normally in cages liberally dusted with DDT. Figure 4–4B shows one way enzymes in resistant flies convert DDT to DDE, which, although harmless to insects, is still toxic to many other species.

That insects can develop resistance to insecticides was demonstrated in the United States as early as 1915, when the increased resistance of scale pests was found to be responsible for the diminishing effectiveness of lime-sulfur sprays and hydrogen cyanide fumigation on West Coast fruit farms in California and Washington. In the early 1940s, the same thing was found in California for bait-sprays of tartar emetic and sugar, first developed in 1939. In South Africa, the blue tick had developed such high resistance to sodium arsenite cattle dips by 1938 that arsenical solutions strong enough to injure the cattle themselves failed to yield effective control (9).

Other Chlorinated Hydrocarbon Insecticides

Thus, entomologists in the late 1940s turned to other chemicals, such as Dieldrin (Fig. 4–5), Lindane (Fig. 4–6B), and Chlordane (Fig. 4–7), all of which initially proved effective against the resistant strains, but whose effectiveness did not last either. With Lindane, 3 months was all it took for new insect strains to develop. These strains were not only unaffected by the Lindane, but were more resistant than ever to DDT, and were also resistant to Dieldrin, even though they had never been exposed to it! Dieldrin, very effective for fly control in 1949, was also found to lose its effect within 3 to 6 months. The use of chemicals similar to DDT (Fig. 4–4C) seemed to renew the vigor of DDT when used with it, at least for 4 to 8 weeks. Nevertheless, the ability of insects to develop resistance to any insecticide was predominant, providing a classic example of Charles Darwin's thesis of survival of (from the insects' point of view) the "fittest." This ability kept scientists busy synthesizing a steady stream of new chemicals, each only temporarily effective. By 1966 the sale of such pesticides had risen to some $500 million per year in the United States alone.

The steady rise in production and use of these pesticides brought a growing realization of their environmental hazards. Sparked by Rachel Carson's timely warning (10), scientists became increasingly aware of the often disastrous side effects of chlorinated hydrocarbons. Paul Ehrlich (1) has listed four especially dangerous properties that these chemicals share:

Fig. 4–5. Chlorinated hydrocarbon insecticides: Aldrin, Dieldrin, and Endrin. Aldrin oxidizes rapidly in Dieldrin, which is isomeric with Endrin. To illustrate the differences between isomers, the carbon atoms have been numbered, and each structure is represented in three dimensions below. In Dieldrin the epoxy portion (carbons 1 and 2) lies to the left of the upper six-membered ring (carbons 4,5,11,10,9), while in Endrin this epoxy portion is directly under that ring.

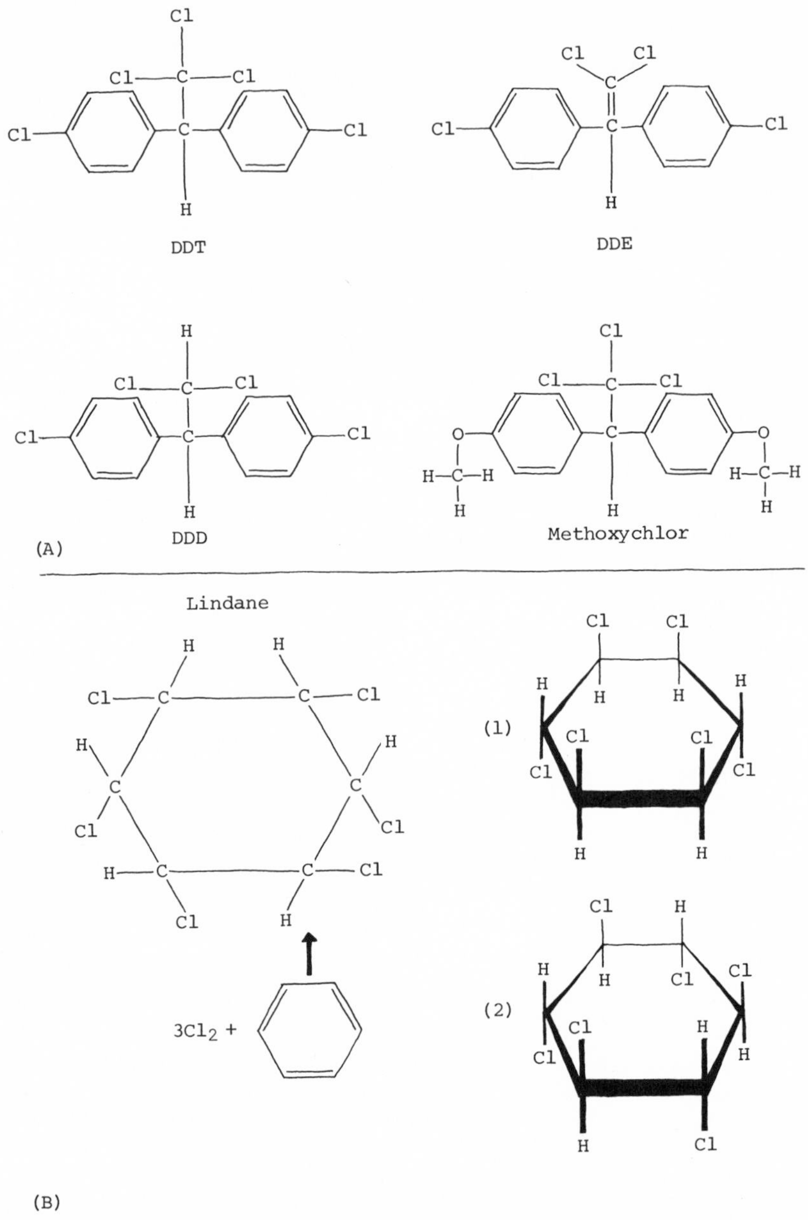

Fig. 4–6. Chlorinated hydrocarbon insecticides: (A) DDT family. (B) Lindane. The synthesis of Lindane (BHC) yields isomers shown at right. Isomer (1) is much more toxic than isomer (2).

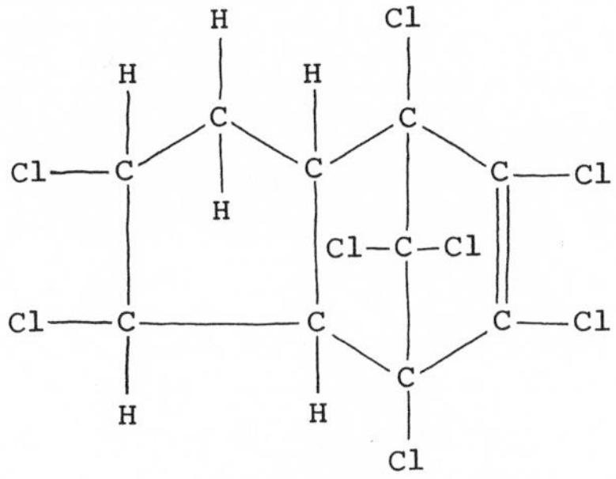

Chlordane

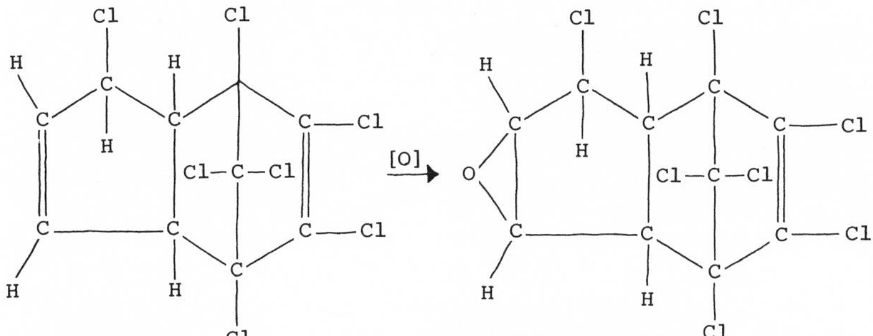

Heptachlor → Heptachlor epoxide

Fig. 4–7. Chlorinated hydrocarbon insecticides: Chlordane family. Chlordane is hard to break down, and hence persists in soil more than twice as long as its relative, Heptachlor, which, because of its reactive double bond, can be converted to its epoxide.

1. First, they are wide-ranging poisons, especially harmful to fish and birds; and possibly to mammals as well.
2. Second, they are very stable. They persist in toxic form for years. In the case of DDT, for example, some 50 percent has been found to remain unchanged over a 10-year period, while much of the remaining 50 percent has been transformed to DDE (Fig. 4–4B). The DDE, still biologically dangerous, may well remain stable for a decade or more.
3. Third, they are mobile, spreading readily and far from where they were initially used. DDT adheres to dust particles, traveling with them around the globe, and evaporates, with water, into the atmosphere. Several different chlorinated hydrocarbons have been found in dust filtered from the air over Barbados, while insects in unsprayed areas high in California's Sierra Nevada mountains have been found to contain DDT.

4. Fourth, these chlorinated hydrocarbons have great affinity for living systems, tending to concentrate in the fat of most organisms. They thus continually move from the physical environment into living things. DDT is so insoluble in water that a saturated solution of it contains only 1 or 2 parts per billion. Most of the DDT washed into our waters ends up in aquatic organisms, in much greater concentrations than this. It is therefore meaningless, as Paul Ehrlich points out, to attempt to monitor DDT levels by testing the water.

The affinity of DDT and its relatives for living things leads to the phenomenon known as biological magnification, in which insecticides are concentrated increasingly from organism to organism, throughout the food chain. The bottom muds in the Green Bay area of Lake Michigan, for example, contained 0.014 ppm of DDT (11). Tiny amphipods living in the mud absorbed the chemical, concentrating it to 0.41 ppm. Fish ate the amphipods, each fish absorbing enough to further concentrate DDT to from 3 to 6 ppm. Herring gulls, each eating many fish, developed DDT levels of almost 100 ppm. This is high enough to interfere severely with the gulls' ability to reproduce, thus eventually threatening their survival.

A marsh along Long Island's south shore, sprayed with DDT for 20 years to control mosquitoes, was found to have DDT residues in its upper layers of mud that reached 32 pounds per acre! At the lowest level of the food chain, the plankton in marsh water contained 0.04 ppm of DDT. Minnows feeding on the plankton concentrated this to 1 ppm, while minnow-eating gulls were found to have about 75 ppm in their tissues (12). This represents almost a 2,000-fold concentration! In 1957 California's Clear Lake contained only 0.02 ppm of DDT, the result of spraying with DDT for gnat control. Microscopic life in the Lake contained 5 ppm, but fish feeding on these organisms contained as much as 2,000 ppm! Fish-eating aquatic birds, such as grebes, died in great numbers. Carnivorous birds suffered the most because their DDT buildups cannot easily be excreted. Western grebes, for example, have been known to store up to 1,600 ppm of DDT residues in their visceral fat (12). The Bald eagle and peregrine falcon populations are bordering on extinction as a result.

The first alarm and eventual explanation of the peregrine falcon's downfall was given by British ecologist Derek Ratcliffe, who found in 1947 that peregrine eggs contained the residues of four insecticides: Lindane, Dieldrin, Heptachlor epoxide, and, most significantly, the DDE formed from DDT. The DDE residues, averaging some 13 ppm, ranged up to over 30 ppm (13). Peregrine eggs had begun to break open prematurely, with fewer and fewer successfully hatched. Ratcliffe also found that peregrine eggshells began to drop in weight by almost 20 percent after the widespread introduction of DDT in the mid-1940s, and that the thinning of the shells was causing the premature breaking. Similar results were obtained in laboratory controlled experiments with mallard ducks upon exposure to DDT.

The emerging explanation is that pesticide irritation of birds' tissues increases liver size, which, in turn, stimulates increased production of a liver enzyme. This enzyme not only helps the bird break down and excrete foreign chemicals, but also breaks down the natural female sex hormones, including estrogens (Fig. 4–8), that are of such critical importance in all stages of the reproductive cycle. When estrogen levels are high, early in the breeding cycle, dietary calcium is stored in the bones. When these estrogen levels fall, normally after mating, the stored calcium is released by the bones into the blood, and carried to the oviduct, where the eggs are made. But if DDT depresses estrogen levels abnormally early in the breeding cycle, too little calcium will be stored in the bones for eggshell production later on. Cornell University scientists have demonstrated that within 24 hours the injection of the DDT residue DDE into a bird inhibits the enzyme carbonic anhydrase, which normally facilitates transport of calcium from the blood to the site of eggshell production in the oviduct, resulting in a 20 to 35 percent reduction in shell thickness (13). Experiments such as these appear finally to have convinced federal government officials to restrict the use of DDT.

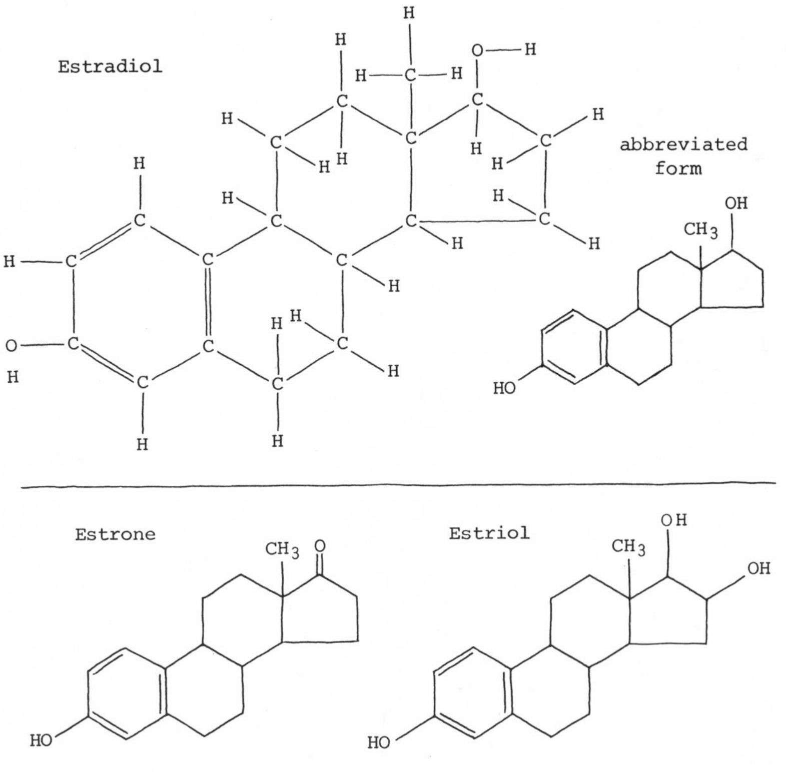

Fig. 4–8. The estrogen female sex hormones, necessary for the normal functioning of the ovulation cycle.

Fish, too, are affected by concentrated DDT residues. Consider the now-famous case of the coho salmon, introduced by biologists into DDT-polluted Lake Michigan in the spring of 1966. For 2 years they thrived and grew fat, while accumulating concentrated DDT residues in their fat from the smaller fish they ate. While large coho don't die of DDT poisoning, some 700,000 newly hatched fish did die in early 1968—in all probability from DDT (11). By the fall of 1968, adult cohos had DDT concentrations of 20 ppm; in early 1969, federal and state officials had to seize some 500,000 pounds of salmon that had been canned or was awaiting processing.

Thus the very properties that initially made DDT such a success have also made it an ecological disaster. Its toxic persistence has carried it up the food chain. Its low solubility in water, but high solubility in fat, has concentrated it and its relatives in the tissues of living things, often with destructive results. While there is little direct evidence that man has been similarly affected, his physiology is much like that of other mammals, so that he, too, stores DDT in his fat, liver, and brain. Studies of 800 samples of human fat from Americans revealed average DDT concentrations of 12 ppm (14). Most of it comes from the food we eat, for we, too, are part of the food chain mentioned above.

Studies of oysters have revealed that during a 7-day exposure to water containing 0.001 ppm of DDT, which is 1 part per billion (ppb), their DDT content reached some 150 ppm! This means a 20-gm oyster can contain 300 μg (0.0003 gm) of DDT, to be passed along to whoever eats the oyster. How does the oyster accumulate such concentrations? By filtering water, 2 liters per hour on the average, or 48 liters per day, 336 liters per week. At 0.001 ppm of DDT, 1,000,000 gm of water contain 0.001 gm of DDT. Therefore, 336 liters (336,000 gm) of water must contain 0.000336 gm of DDT. The oyster studies thus showed that over 90 percent of this DDT, 0.0003 gm of it, remained in the little animals (15).

Finally, the chlorinated hydrocarbon insecticides' toxic effects on a wide variety of organisms, and their ability to travel around the globe, far from sites of original application, make them a deadly threat to all life on Earth. Even the ocean is in danger, because as dilute as DDT and its relatives may be in the waters of the ocean, marine life concentrates them with amazing efficiency, while their persistence keeps them toxic for many years. Even if the United States stops using DDT, its continued use in other countries would have a devastating worldwide effect. Should worldwide use of DDT be stopped tomorrow, its direct effects would continue to be felt for 10 to 20 years, while its indirect effects—the decimation or extinction of certain species—may be felt for centuries. The lesson is obvious: Simply because we can produce a chemical that will be of immediate benefit by killing pests, we *must* learn of its long-term effects *before* allowing its widespread use.

Organophosphate Insecticides and Herbicides

The organophosphate insecticides are all descended from a nerve gas, diisopropylfluorophosphate, developed by Nazi Germany in World War II. A

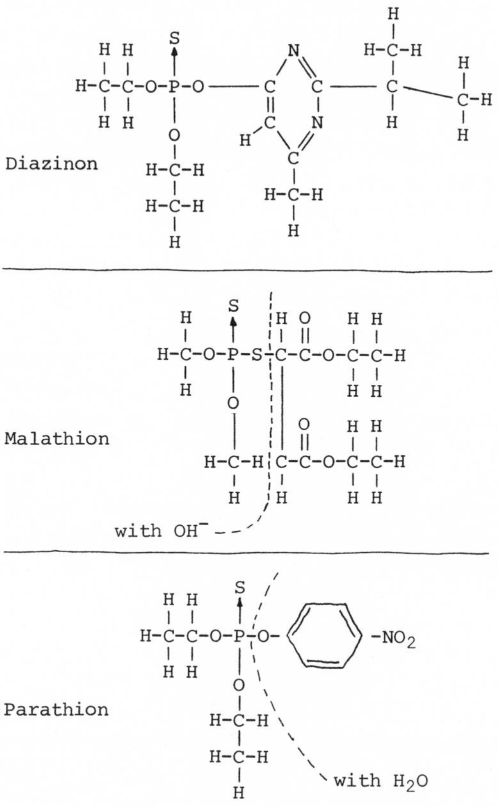

Diazinon

Malathion

with OH⁻

Parathion   –NO₂

with H₂O

Fig. 4–9. Organophosphate insecticides. Dotted lines show where Malathion and Parathion can be split into small, more biodegradable fragments, which is thought to account, in part, for their lack of persistence in soils.

look at some of their structures (Fig. 4–9) reveals the basic phosphate (PO_4^{3-}) group, but with one or more of the oxygens replaced by sulfur atoms. Phosphorus, with its five valence electrons, usually forms three normal covalent bonds, as in phosphine (PH_3), the phosphorus analog of ammonia (NH_3). The two so-called 3s electrons (Fig. 4–10 top right) often also form a coordinate or dative bond, shown as an arrow ($\longrightarrow$), with the phosphorus atom contributing both electrons to the bond. Such a bond is shown forming, in Fig. 4–10, between the oxygen atom at left, and the phosphorus atom, and makes up one of the bonds in phosphoric acid (H_3PO_4). This acid can ionize in three stages: to dihydrogen phosphate ($H_2PO_4^-$), monohydrogen phosphate (HPO_4^{--}), and phosphate ion (PO_4^{3-}). The first two ions are found in blood and living tissue.

(A) Phosphorus, showing bonding electrons, can form three normal covalent bonds with its three 3p electrons, and one coordinate, or dative bond, by furnishing both 3s electrons. Both types are exhibited in the formation of phosphoric acid:

orbital diagram

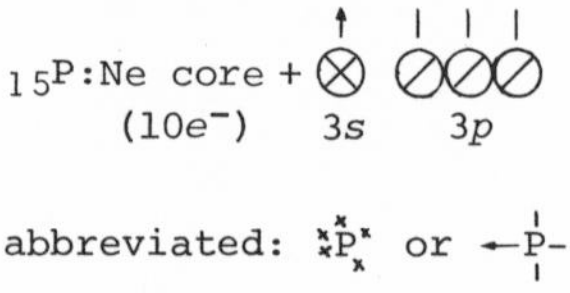

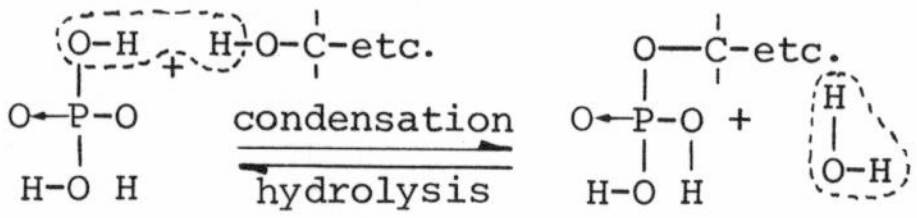

$_{15}$P:Ne core + ⊗ ⊘⊘⊘
(10e^-)   3s 3p

abbreviated: ⁚P⁚ or —P—

$$:\ddot{O} + :P: + .\ddot{O}:H \longrightarrow :\ddot{O}:P:\ddot{O}:H \quad or \quad O—P—O—H$$

Pyrophosphoric Acid

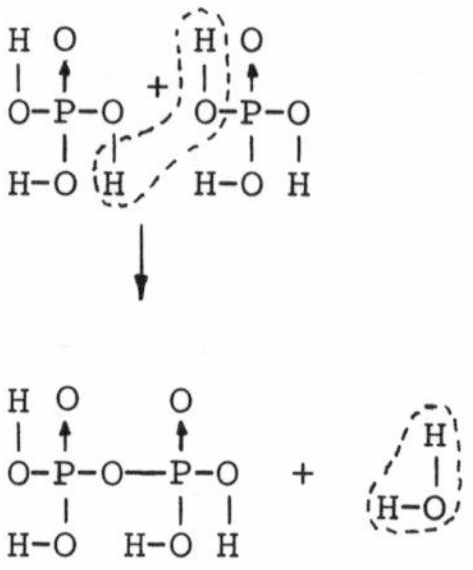

(B) Phosphoric acid (H$_3$PO$_4$) can react with the OH groups of sugars (above) as in DNA portion shown at right; or with itself to form pyrophosphoric acid (left), or ATP (adenosine triphosphate, below).

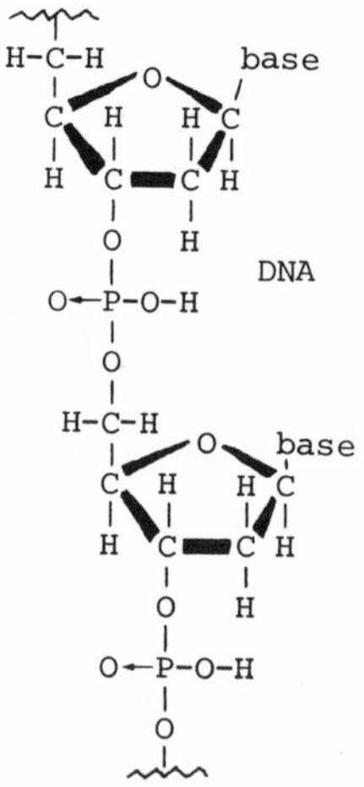

DNA

ATP

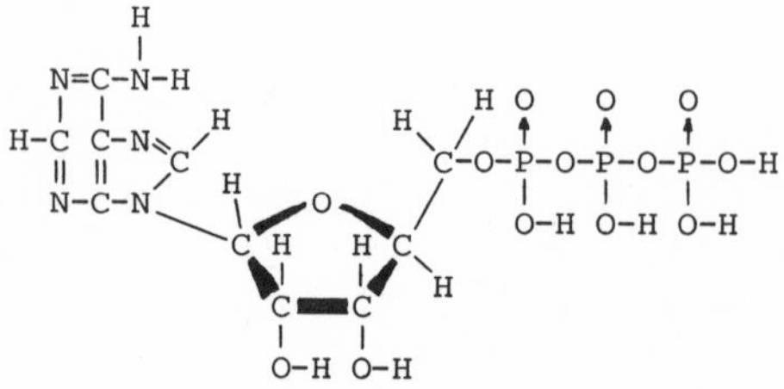

Fig. 4–10. Basic chemistry of phosphorus.

Typical reactions of phosphates include condensations, in which they combine with the OH groups of organic alcohols or carbohydrates by splitting, or condensing out a molecule of water (Fig. 4–10). They can also combine with each other by condensing out water between their own OH groups. The reverse of such reactions, where the addition of water results in breaking down the condensation product into the two smaller original molecules, is known as *hydrolysis*. Hydrolysis can also be induced in some phosphate compounds by hydroxide (OH^-) ions (see Fig. 4–9). Note that by condensing out water from deoxyribose-type sugars, phosphate groups link them together to form the genetic information-carrying DNA molecules present in living cells (Fig. 4–10). And, by condensation among themselves, phosphoric acid molecules form an important part of the ATP molecules so essential in biological oxidation processes.

The organophosphate insecticides all inactivate the enzyme cholinesterase, which, in turn, normally breaks down the so-called nerve "transmitter substance," acetylcholine. When acetylcholine is not broken down, because of inactivated cholinesterase, the insect dies of hyperactivity of the nervous system. Because organophosphates can easily be broken down by hydrolysis, compounds such as Malathion and Parathion (Fig. 4–9) tend to disappear from the environment in a week or so, and do not, therefore, tend to build up in living tissues. This is in marked contrast to the extremely stable, persistent chlorinated hydrocarbons such as DDT. These phosphate insecticides can *phosphorylate*, or break down, enzymes and other parts of living systems. Indeed, their ability to thus poison an esterase enzyme, which is more critical to insects than to mammals, is thought to account in part for their selective ability to kill insects while not harming mammals. Malathion, an effective insect killer, is relatively non-toxic to mammals for another reason: Mammals contain the enzyme, carboxyesterase, which destroys Malathion. If, however, Malathion is used together with other organophosphates, the combination seems to inhibit mammals' carboxyesterase, and is toxic to them.

Parathion, used since the late 1960s in increasing amounts, has come to be known as DDT's successor. Even though its rapid decomposition is a long-term advantage, its original form is some 300 times more toxic than DDT. It has been labeled by Dr. John Davies of the University of Miami School of Medicine as the leading insecticide responsible for human poisonings and death since 1966 (16). Indeed, it is chemically similar to the nerve gas that has become such a storage problem for the U.S. Army. Dangerous for several weeks after spraying, Parathion caused six of the nine pesticide poisonings in Dade County, Florida, in 1970, including two fatalities. It also killed a 7-year-old in North Carolina, almost killed his 11-year-old brother, and built up such concentrations in other members of the family that one more exposure to their sprayed tobacco field might have been fatal—even though the family was reported to have waited longer than the recommended 5 days after spraying before entering the field. North Carolina doctors were puzzled by the special susceptibility of children to Parathion, speculating they may not have been able to build up tolerances to it. Readily absorbed through the skin, it attacks the nervous system by inhibiting the nerve-impulse-controlling enzyme cholinesterase. As a result, body movements become convulsive and uncoordinated. First symptoms are headaches, dizziness, nausea, and breathing difficulties, followed by tremors, convulsions, and paralysis.

Most pesticides are reactive chemicals, which accounts for their effectiveness. We have seen, for example, that Parathion kills because it reacts with the active sites of acetylcholinesterase enzymes, and that it undergoes hydrolysis, which is typical of phosphate compounds. In all probability, however, only a small portion of an applied pesticide may contact the target organism. Most of it, unfortunately, will undergo chemical transformations by reaction with the outside environment, rather than through plant or animal metabolism. According to Donald G. Crosby (17), light, especially ultraviolet light of wavelengths less than 0.4 micron, is most important in initiating many so-called photochemical reactions. Sunlight alone apparently can do this, for grass treated with Dieldrin has been found, after being exposed to ordinary daylight, to contain a chlorinated transformation product, an isomer of Dieldrin, in which carbon atom 7 (C-7) is bonded to either C-9 or C-10, the latter atoms thus connected only by a single bond (see Fig. 4–5). This isomer was reported to be twice as toxic as the original Dieldrin to insects, and as much as four times as toxic to mice.

Aldrin and Dieldrin, even in light below the atmosphere's ultraviolet limit, undergo a change by losing the chlorine atom on C-9 or C-10, replacing it with a hydrogen atom. This compound is about four times more toxic orally to mice than is the parent insecticide (17). The irradiation of Parathion, by either sunlight or ultraviolet light in the laboratory, produces compounds more polar than the Parathion, but apparently less toxic. Additional oxidation produces a more toxic oxygen analog, in which the sulfur (S) atom is replaced by an oxygen (O) atom. The sulfur-containing side chains found in organophosphates are prime targets for photo-oxidation, often undergoing the type of reaction just described.

Water, because of its ability to form both hydroxyl (OH) and hydrogen (H) radicals and ions, often enters into reaction with pesticides, the hydrolysis reaction (Fig. 4–10) being the most common. If the water is alkaline, with pH above 7, Malathion is hydrolyzed to the two fragments indicated in Fig. 4–9, the right-hand fragment forming a double bond between the carbon atoms to the left of the C=O groups. Soils, especially those containing acidic clays, also initiate reactions. Endrin, for example, reacts to form a ketone, where the oxygen bonded to C-1 and C-2 (Fig. 4–5) double-bonds to either C-1 or C-2 (17).

Herbicides such as 2,4-D, 2,4,5-T, and Picloram (Fig. 4–11), have been widely used recently in place of mechanical means for aiding crop cultivation, keeping railway roadbeds and roadways free of shrubs, and for military defoliants in Vietnam. These compounds are related to the growth-regulating hormone, indoleacetic acid, and cause uncontrolled growth and metabolism, and, eventually, death. Because they do not act as growth substances in animals, they do not harm them directly, although increasing evidence indicates that 2,4-D may constitute a potential danger to fish even in normal use. Its rate of use in Vietnam has been greater than this. By 1969 scientific studies gave strong evidence that 2,4-D and 2,4,5-T caused birth malformations in animals, leading in the spring of 1970 to their curtailment in the United States and in Vietnam.

The defoliant program in Vietnam resulted, during the 1960s, in the spraying of an estimated 40 million pounds of 2,4,5-T, as well as lesser amounts

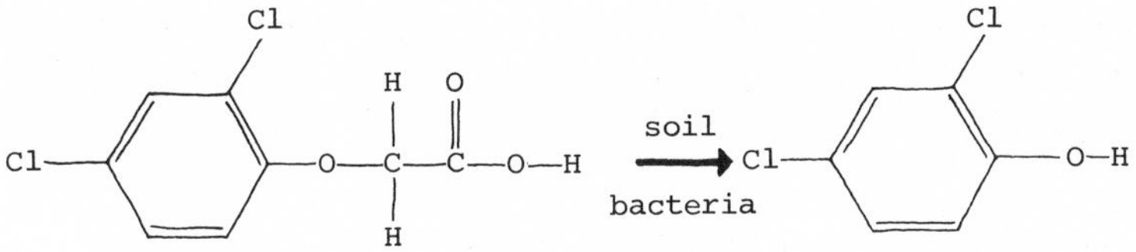

2,4-D

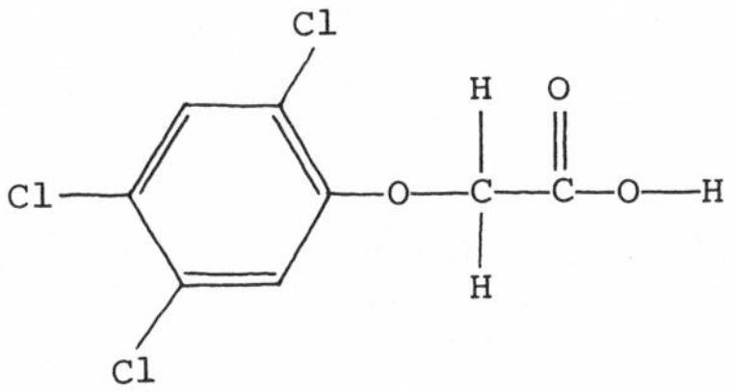

2,4,5-T

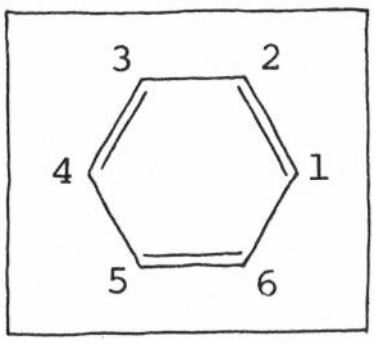

Numbered positions in benzene ring (box above) show how the numbers in 2,4-D and 2,4,5-T indicate where in the ring the chlorine atoms are attached.

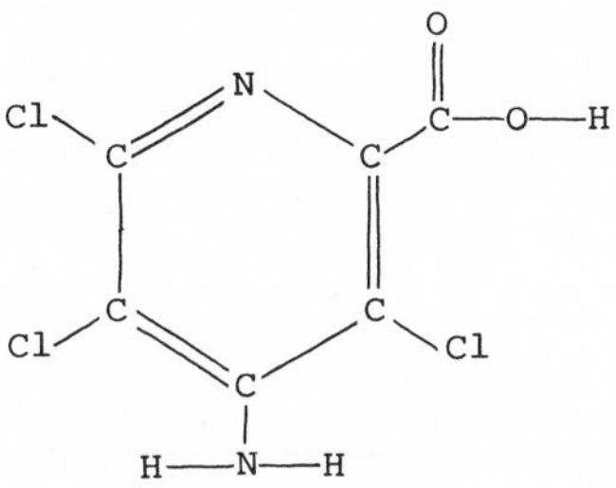

Picloram

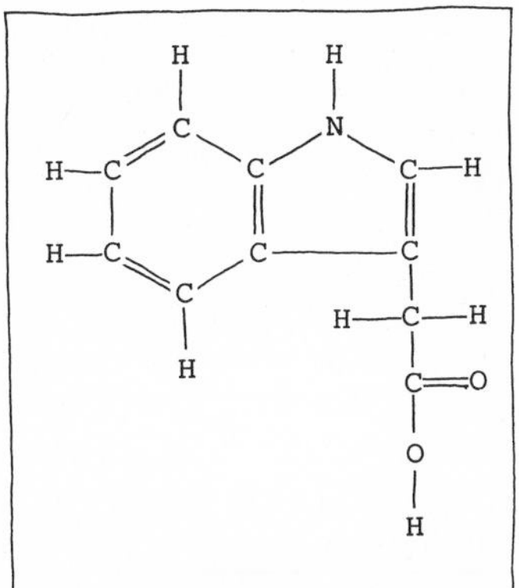

Indoleacetic acid

Fig. 4–11. Herbicides: 2,4-D and the less toxic product it is converted to in soil; 2,4,5-T; and Picloram.

of 2,4-D and Picloram, across at least 5 million acres. While 2,4-D is rapidly metabolized in woody plants and effectively oxidized by non-living environmental systems (the light, air, water, and soils already mentioned), 2,4,5-T remains active somewhat longer. Picloram can remain active for 18 months or more (see Table 4–1), and can thus circulate in the environment in toxic form. In addition, Picloram is more active than the other herbicides, and has not,

Table 4-1. Major Pesticides in Use Since 1940

Type	Chemical name	Persistence in soil*
Chlorinated hydrocarbons		
Aldrin	1,2,3,4,10,10-hexachloro-1,4-4a,5,8,8a-hexahydro-1,4-*endo-exo*-5,8-dimethano-naphthalene	2 years
Chlordane	1,2,4,5,6,7,8,8-octachloro-2,3,3a,4,7,7a-hexahydro-4,7-methanoindane	5 years
DDD (TDE)	*d*ichloro*d*iphenyl*d*ichloroethylene	
DDT	*d*ichloro*d*iphenyl*t*richloroethane	4 years†
Dieldrin	1,2,3,4,10,10-hexachloro-6,7-epoxy-1,4, 4a,5,6,7,8,8a-octahydro-1,4-*endo-exo*-5, 8-dimethanonaphthalene	3 years
Endrin	1,2,3,4,10,10-hexachloro-6,7-epoxy-1,4, 4a,5,6,7,8,8a-octahydro-1,4-*endo-endo*-5,8-dimethanonaphthalene	3 years‡
Heptachlor	1,4,5,6,7,8,8-heptachloro-3a,4,7,7a-tetra-hydro-4,7-endomethanoindene	2 years
Lindane (BHC)	1,2,3,4,5,6-hexachlorocyclohexane	3 years
Methoxychlor	Dimethoxydiphenyltrichloroethane	
Organophosphates		
Diazinon	*O,O*-diethyl-*O*-(2-isopropyl-4-methyl-6-pyrimidyl)phosphorothioate	12 weeks
Malathion	*O,O*-dimethyl-*S*-(1,2-dicarbethoxyethyl)-phosphorodithioate	1 week
Parathion	*O,O*-diethyl-*O*-*p*-nitrophenylphosphoro-thioate	1 week
Herbicides		
2,4-D	2,4-dichlorophenoxyacetic acid	1 month
2,4,5-T	2,4,5-trichlorophenoxyacetic acid	5 months
Picloram	4-amino-3,5,6-trichloropicolinic acid	18 months

*Time for 75% to disappear from soil. From *Chemical Fallout*, Miller and Berg, Eds. Charles C. Thomas, Springfield, Ill., 1969, pp. 54–61.
†Misleading since DDT concentrates in living organisms and remains as DDT for 10 years or more.
‡Dieldrin isomer.

therefore, been licensed by the Federal Drug Administration for use on American crops. Because of its persistence, it can be dissolved out of areas where it was applied, flowing to unsprayed regions where it has been known to injure crops severely as much as a year after original application (18). Studies of 2,4-D have shown it to be metabolized to a variety of products by soil microflora and plants, and to a lesser extent by mammals, resulting in products more toxic to mammals than the 2,4-D itself. It seems clear that considerably more work is needed before we can predict the ecological effects of introducing synthetic pesticides and herbicides.

Discovery of Radioactivity

The fallout, and subsequent concentration, of the long-lived chlorinated hydrocarbon insecticides such as DDT is similar to the fallout of another class of pollutants which also began to be widely distributed following World War II, radioactive substances. While natural radioactivity is as old as the universe, additional exposure from artificially induced radioactivity began in 1896 with the discovery and subsequent use of x-rays, and became widespread after the development of atomic weapons in the mid-1940s.

Natural radioactivity was first discovered by the French scientist Henri Becquerel in 1896, and was explained six years later by the British physicists Rutherford and Soddy. All elements with atomic numbers greater than 82 are naturally radioactive. Natural uranium, for example, atomic number 92, spontaneously decays by emitting alpha-particles. An alpha-particle is simply a helium nucleus, consisting of 2 protons and 2 neutrons. Natural uranium, more than 99 percent uranium-238 (U-238 for short), is made of atoms whose nuclei each contain 92 protons and 146 neutrons, a total of 238 nuclear particles (called *nucleons*). As indicated in Fig. 4—12A, the emission of an alpha-particle leaves a nucleus containing 2 less protons and 2 less neutrons, so that it is no longer uranium but an isotope of thorium. This type of radioactivity, called alpha-emission, always results in a new substance whose atoms are lighter by the weight of the emitted alpha-particle.

The thorium thus produced is also radioactive, but it is a beta-emitter, spontaneously giving off beta-particles, or electrons, from the nucleus. Beta-decay always results in formation of the next heavier element, a new substance of the same weight, but containing 1 more proton and 1 less neutron than previously (Fig. 4—12A). Such beta-activity can thus be explained by imagining that a nucleus with too many neutrons becomes more stable by a process in which 1 neutron emits an electron (the beta-particle) to become a proton. The half-lives for each process, written on the equation arrows, are strikingly different: 4½ billion years for half the uranium-238 in a sample to decay by alpha-emission to thorium-234, only 3½ weeks for half the thorium thus produced to disintegrate by beta-decay to protactinium-234. The protactinium (Pa) is also radioactive, undergoing some dozen additional radioactive disintegrations before ending up as an isotope of lead, stable lead-206. Radium-226, the most abundant isotope of radium, is but one radioactive disintegration product.

One of the first instances of the harmful effects of radiation was the radiation poisoning of workers who applied radium paint to watch-dial numbers

$$\begin{array}{c}
^{238}_{92}U \xrightarrow[\text{years}]{4.5 \times 10^9} {}^{4}_{2}He + {}^{234}_{90}Th \xrightarrow[\text{days}]{24.1} {}^{0}_{-1}e + {}^{234}_{91}Pa
\end{array}$$

Uranium-238   Thorium-234   Protactinium-234

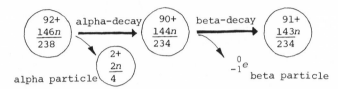

(A) Natural Radioactivity. Uranium-238 spontaneously decays by alpha-emission into a thorium isotope (top left), which in turn undergoes beta-decay to become protactinium (top right) Half-lives are shown on equation arrows.

$$^{1}_{0}n + {}^{238}_{92}U \longrightarrow {}^{239}_{92}U \xrightarrow{\;{}^{0}_{-1}e\;} {}^{239}_{93}Np \xrightarrow{\;{}^{0}_{-1}e\;} {}^{239}_{94}Pu$$

Uranium-238   Neptunium-239   Plutonium-239

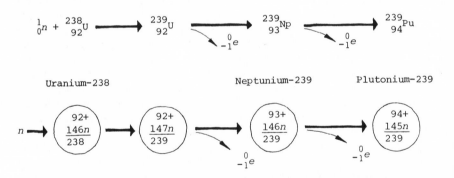

(B) Neutron bombardment makes uranium-238 unstable (too many neutrons), and by undergoing beta-decay it reduces the number of neutrons. This is accompanied by an increase in the number of protons, creating new heavy elements.

Fig. 4–12. Nuclear reactions: (A) Natural radioactivity. (B) Neutron bombardment.

to make them luminous, and who years later developed fatal cancers from this exposure. One study of miners in the radium-ore-rich Schneeberg district of East Germany established that exposure to the radioactive ore caused fatal lung cancer in from 60 to 70 percent of Schneeberg's miners.

Prior to the discovery of the neutron in 1932, much nuclear research involved bombarding atomic nuclei with high-energy alpha-particles from uranium or radium, or using electricity to accelerate other positive particles which could then be directed against these nuclei. Just as the internal structure of a building is revealed by breaking away its outer walls, so physicists hoped to gain an understanding of nuclear structure by bombarding the nuclei of atoms. Because of the electrostatic repulsion of positive particles by the positive nuclei, the bombarding particles had to be given tremendous energies to enable them to overcome this repulsion. Italian physicist Enrico Fermi reasoned that neutral neutrons would experience no such repulsion, and so would make ideal bombarding particles. Having no motive other than scientific curiousity, Fermi and his associates accordingly bombarded dozens of heavy elements with neutrons, and found that they absorbed the neutrons and became radioactive beta-emitters, spontaneously changing to the next heavier element. This could be explained by assuming that these artificially created elements tried to reduce their neutron populations by changing a neutron to a proton through emission of an electron (beta-particle).

If bombarding heavy elements with neutrons produced still heavier elements, it was natural for Fermi to try the same technique with uranium, the heaviest element then known. If it followed the pattern, it should transmute into a totally new, even heavier element. The neutron bombardment of uranium did produce new elements: element 93 (neptunium), as well as element 94 (plutonium) (Fig. 4–12B). More important, however, was the completely unexpected behavior of uranium-235, the lighter isotope that makes up only 0.7 percent of all natural uranium. It absorbs a neutron and splits, or fissions, into two approximately equal fragments, releasing more neutrons as well as vast amounts of energy. These newly released neutrons were capable of causing other U-235 atoms to fission, releasing more energy and more neutrons, which, in turn, could sustain a so-called chain reaction, resulting in the release of huge amounts of energy. The practical effects of this process were first seen a decade later, with the explosion of the first so-called atomic bombs—made of uranium-235 and plutonium-239, which also fissions when absorbing neutrons.

Radioactive Fallout from Atomic Bomb Tests

Once separated, pure U-235 or Pu-239 will undergo spontaneous fission chain reactions and explode, providing they are above a certain *critical size*. At less than this size, neutrons released in initial fissions simply wander out of the "bomb" before they can strike other uranium and plutonium nuclei to keep the fission chain alive. These fission bombs are the type we dropped on the Japanese cities of Hiroshima and Nagasaki, and which the U.S. and U.S.S.R. frequently tested in the atmosphere prior to the signing of the limited nuclear test ban treaty in late 1963. (Although the French and Chinese still conduct occasional atmospheric tests, we and the Soviets now test underground.)

As shown in Fig. 4–13A, each fission bomb produces many different fission fragments; one uranium atom may split into barium and krypton, another into cesium and rubidium, another into xenon and strontium, another into iodine and yttrium, and so on. Each such fragment is radioactive, having many more neutrons than the normally occurring stable isotopes found in nature. Most normal iodine atoms, for example, contain 74 neutrons in their nuclei, while the radioactive iodine-131 formed in atomic bomb explosions has 78 neutrons. They therefore tend to emit beta-radiation which, as we have seen, is accompanied by neutrons apparently "changing" into protons. Strontium-90, a beta-emitter, decays to an isotope of yttrium:

$$^{90}_{38}Sr \longrightarrow ^{\ 0}_{-1}e + ^{90}_{39}Y$$

These fragments make up the radioactive fallout accompanying atom bomb tests. Three of the fission products shown in Fig. 4–13A are especially dangerous to man because they are intimately connected with his biochemistry. These are strontium-90 (Sr-90), comprising some 5 percent of the total atoms produced in fission, cesium-137 (Cs-137), and iodine-131 (I-131). Strontium is chemically similar to calcium, and therefore readily absorbed by living things, passing from the soil and vegetation to animals who eat the vegetation, and finally to man, the vegetation-eater, the milk-drinker, the flesh-eater. Strontium-90 has a half-life of about 28 years, so that when it deposits in man, primarily in the bones and bone marrow, it can contribute to skeletal radiation for a lifetime. Strontium-89, also produced in fission, has a much shorter half-life (only 51 days), so that its radioactivity dissipates after a year or two, making it especially dangerous to young children who use most of the calcium (and hence strontium) in their diet to form the skeleton. Short-lived isotopes are initially dangerous because they undergo most of their radioactive disintegrations quickly, soon becoming relatively inactive. The longer-lived varieties undergo far fewer disintegrations per unit of time, but maintain a steady output for years. A single microgram of Sr-89, for example, undergoes 45 billion disintegrations per minute during the first 51 days of its existence, making it over 200 times as active as an equivalent amount of Sr-90. After 2 years, however, while the activity of Sr-90 remains undiminished, Sr-89 is 20,000 times less active than initially, making it 100 times less active than Sr-90. Every disintegration is potentially dangerous to living cells, so that long-lived isotopes that tend to store in the body can be as dangerous in the long run as initially reactive, short-lived species. Animal experiments have shown that both Sr-89 and Sr-90 can produce leukemia, bone cancer, and other skeletal effects (18).

Cesium-137 is formed in slightly greater amounts that strontium-90 (about 6.2 percent of total fission products), and has a similar half-life, some 27 years. It emits both beta-particles and gamma-rays (similar to high-energy x-rays). While it does not become fixed in the body, it does make a major contribution to long-lived and highly penetrating gamma-activity. As such, its

Fig. 4–13. Nuclear reactions accompanying atomic weapons development: (A) Atomic bomb reaction. (B) Hydrogen bomb reaction.

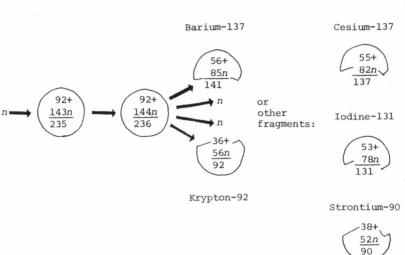

$$_{0}^{1}n + {}_{92}^{235}U \longrightarrow {}_{92}^{236}U \rightarrow \text{neutrons + radioactive fission products}$$

Barium-137

Cesium-137

$n \longrightarrow$ 92+ / 143n / 235 $\longrightarrow$ 92+ / 144n / 236

56+ / 85n / 141

n

n

36+ / 56n / 92

or
other
fragments:

55+ / 82n / 137

Iodine-131

53+ / 78n / 131

Krypton-92

Strontium-90

38+ / 52n / 90

(A) Atomic bomb reaction, in which pure U-235 fissions into two
fragments whose atomic numbers add up to 92. Several fragments
most dangerous to human health are shown at right; they represent some 14 percent of all fission products.

$${}_{3}^{6}Li + {}_{0}^{1}n \rightarrow {}_{1}^{3}H + {}_{2}^{4}He$$

$${}_{1}^{2}H + {}_{1}^{3}H \rightarrow {}_{2}^{4}He + {}_{0}^{1}n$$

$${}_{0}^{1}n + {}_{7}^{14}N \longrightarrow {}_{6}^{14}C + {}_{1}^{1}H$$
Atmospheric

$${}_{3}^{6}Li{}_{1}^{2}H$$ A-bomb

(B) Hydrogen bomb (top right) has atom bomb surrounded by
lithium-6 deuteride. Atom bomb furnishes neutrons and
high temperatures which initiate fusion reactions (left),
which release neutrons to react as shown in the final
equation.

principal potential hazard is genetic, from irradiation of gonadal tissue. Iodine-131 makes up some 3 percent of fission products, but has a half-life of only 8 days, making it extremely dangerous for several weeks following its formation. Within 2 to 4 days after being deposited on pastures and consumed by grazing dairy cattle, levels of I-131 in their milk reach a maximum. The iodine from milk thus contaminated and drunk by humans concentrates in the thyroid glands. This is especially hazardous to young children and infants, who consume large quantities of milk, and whose thyroids are only one-tenth the size of adult glands. The small size of childrens' thyroids is significant because the total radiation dose to the gland (measured in units called *rads*) is proportional to the amount of I-131 per gram of thyroid. Adult thyroids can receive at least 4,000 rads of I-131 with no apparent damage, but strong evidence indicates that 200-rad doses will cause thyroid cancer in 3 percent of exposed young children. The Federal Radiation Council notes that thyroid cancers have occurred in children after exposures even lower than this.

Gamma-radiation is much more penetrating than alpha-radiation, which is stopped by clothing, or beta-radiation, which is somewhat more penetrating than alpha-radiation. Thus the half-dozen short-lived, initially dangerous gamma-emitting fission products from atom bomb tests can cause hazardous whole-body exposure without entering the body itself. These and other radioactive fission products can also enter the body through inhalation, where they can irradiate lungs and be absorbed into the circulatory system, there to be carried to critical tissues and organs.

Fallout from atom bomb tests falls into three categories. Relatively large, heavy radioactive particles (larger than 50 microns) are deposited within several days, a hundred miles or so downwind of the test site. High concentrations of short-lived radioactive products, deposited over a limited area, make this local fallout intensely radioactive. A second portion of radioactive debris enters the *troposphere* (between 6 and 12 miles above the Earth), and is normally confined to a relatively narrow zone of latitude. Most falls to Earth within a few weeks after its formation. Tropospheric fallout from the U.S. Atomic Energy Commission's Nevada test site has caused high levels of radiation several hundred miles away. The remaining debris enters the stratosphere, where it may take years for the bulk of it to be deposited on the ground. Although this fallout is worldwide, most of it will come down in the hemisphere where it was injected. The global range and effects of radioactive fallout allow scientists to determine whether or not an atomic bomb has been exploded above ground anywhere in the world, permitting both the U.S. and U.S.S.R. to monitor each other's observance of the limited test ban treaty without setting foot in the other's territory.

The Hydrogen Bomb

When it became apparent in the late 1940s that a hydrogen bomb was technically feasible, opponents such as J. Robert Oppenheimer were overruled and the bomb was developed. It utilizes essentially the same reactions that occur in the Sun and other stars, whereby hydrogen is fused together to form helium, accompanied by a loss of mass and the release of staggering amounts of energy, in accordance with Einstein's equation, $E = mc^2$ (energy equals mass times the velocity of light squared). Deuterium ($_1^2$H) and tritium ($_1^3$H) will fuse at the high temperatures produced in an atomic bomb, and lithium-6 will react with neutrons to produce tritium (Fig. 4—13B). The most straightforward procedure, therefore, would be to surround an ordinary atomic bomb with a coating of lithium hydride (LiH), a salt-like material, using lithium-6 and deuterium to make the hydride. This coating of $_3^6$Li$_1^2$H would then be available so that the $_3^6$Li could react with neutrons produced in the atomic bomb explosion to form the $_1^3$H, which would then fuse, when triggered by the high temperatures of the atom bomb, with the $_1^2$H of the hydride. This so-called H-bomb reaction also produces neutrons, which, along with those from the atom bomb explosion, bombard the nitrogen of the atmosphere to produce carbon-14 (Fig. 4—13B).

An atomic bomb is obviously limited in size and therefore in explosive force; each of its separate components cannot exceed the critical size. When brought together to create a whole bomb greater than critical size, the thing fissions spontaneously. Theoretically, hydrogen bombs have no such limitation because larger amounts of $_3^6$Li$_1^2$H can be used to surround the atom bomb trigger. While the Hiroshima atomic bomb, dropped on August 6, 1945, was reported to have an explosive force equivalent to 20,000 tons of TNT (20 *kilotons*), the famous Eniwetok H-bomb test of March 1, 1954, produced a destructive force of 12 to 14 *megatons* (1 megaton is equivalent to 1 million tons of TNT). The Soviet Union reportedly exploded hydrogen bombs of over 50 megatons during the late 1950s. The 1954 Eniwetok test was conceded by the Atomic Energy Commission to be twice as powerful as anticipated, and a shift in a sea breeze dumped radioactive ash from the explosion on the Japanese fishing boat *Lucky Dragon*, even though the boat was 120 miles away and well out of the so-called "danger" zone. The crew was seriously injured, one member subsequently dying from the effects of radiation.

From 1945 to 1962, a total of 423 nuclear tests was announced by the nuclear powers, the U.S. held 271; Britain, 23; France, 5; and the U.S.S.R., 124. The total yield from these tests, excluding the 60-odd tests in Nevada, was some 511 megatons, of which 193 megatons was from atomic (fission) bombs (19). Besides the radioactive fission products, the tremendous quantities of neutrons from the hydrogen bombs irradiated surface materials and produced carbon-14 by bombarding atmospheric nitrogen (Fig. 4—13B). Carbon-14 is a weak beta-emitter, with a long half-life of almost 5,600 years. Carbon, however, is a basic element of living matter, so that any radioactive carbon-14 will readily be taken into the body and deposited throughout the system—not only irradiating the entire body, but being passed along to future generations via the chromosomes of the irradiated parents. Nuclear explosions produce carbon-14 that adds significantly to the carbon-14 normally present in nature. In mid-1962 the carbon-14 content of the atmosphere was estimated to have increased 37 percent since testing began; by mid-1963 this figure had jumped to 80 percent.

Underground Nuclear Tests

Although the United States conducted underground atomic bomb tests as early as 1957, most such tests have been conducted at the Nevada test site following the 1963 treaty. As of mid-1969 the Atomic Energy Commission

(A.E.C.) had admitted to 200 underground tests, while the Soviet Union had conducted 50 at their Semipalatinsk and Novaya Semlaya sites in Siberia. The British and French have conducted a few in Australia and the Sahara Desert, respectively. Most of the nuclear explosives that have been or will be tested consist of the fission atomic bomb trigger and the thermonuclear, or hydrogen bomb-type portion, so that besides the fission products, radioactive tritium and large numbers of neutrons are produced. Although most of the tritium is presumably fused (as shown in Fig. 4—13B), the intense neutron bombardment often produces tritium by reaction with lithium in the surrounding minerals, irradiating the bomb casing as well as other materials at the explosion site.

The radiation thus released undoubtedly contaminates local groundwater supplies. Tritium especially, being chemically identical to hydrogen, quickly becomes incorporated into surrounding groundwater, moving where the water does. As long as the groundwater percolates slowly through earth and rock, tritium's half-life of 12 years will ensure its becoming relatively inactive before it reaches human water supplies. The trouble is that the Nevada test site has many natural fissures, as well as those created by the tests themselves, and groundwater is known to flow much faster along fissures than through granular rock. The present monitoring of water supplies along known faults and fissures leaves much to be desired.

Besides the underground radiation, underground tests occasionally release radiation into the atmosphere. The A.E.C. has admitted to fourteen of these so-called "ventings" prior to mid-1968, some releasing as much radiation to the air as an atmospheric explosion the size of the Hiroshima bomb. Such ventings occur when the test explosion forces radioactive gases to escape through the hole containing the bomb, or through natural cracks and fissures in the rock as well as those created by the explosion. This radiation is usually local, but may therefore be concentrated enough to be hazardous to nearby communities. Several scientists believe that venting was responsible for the unusually high iodine fallout recorded in several western states in the early 1960s (19).

The Atomic Energy Commission also has a program known as Project Plowshare, designed to develop peaceful uses of nuclear explosives. In operation since 1957, it has explored the use of nuclear blasts to build harbors, dams, highway cuts, and canals, and to stimulate oil and natural gas production by following up the widely used practice of detonating ordinary chemical explosives in oil- and gas-bearing strata. This practice, used with varying degrees of success, is now being tried on a much larger scale. Nuclear explosives are being used to fracture large volumes of rock, hopefully freeing trapped gas or oil. The problem is that all such explosions produce radioactivity, and a growing number of scientists have become concerned that the dangers of increased radioactive pollution will outweigh whatever benefits Plowshare might produce. The A.E.C. has been less than frank in owning up to some of these dangers, for the use of nuclear explosives permits a vast amount of energy to be packed in a very small space, such as a well, making the technique highly attractive to the oil and gas industries.

The first test of this technique, known as Project Gasbuggy, took place 4,240 feet below ground in a desolate area of New Mexico known as the San

Juan Basin, on December 10, 1967. A 26-kiloton (equivalent to 26,000 tons of TNT) nuclear "device" was exploded in a sealed well. Just as the A.E.C. had promised, no radioactivity was released to the atmosphere at the time of the blast. But, just 1 month later the hole was opened, and large quantities of radioactivity produced in the explosion were deliberately released into the environment. Much of the radioactivity is in the natural gas itself, a mixture of methane (CH_4) and propane (C_3H_8) made far too radioactive by the explosion for normal use. Accordingly, it must be disposed of, and the most convenient way of doing so is by *flaring*, i.e., burning it in the open air at the entrance to the well. The radioactive natural gas, along with most of the gaseous radioactive by-products of the explosion, is thus released into the air, inert krypton isotopes passing out unchanged, while the tritium (radioactive hydrogen-3) and natural gas are burned to give radioactive carbon dioxide and radioactive water vapor. Thus Project Gasbuggy, run jointly by the Atomic Energy Commission and the El Paso Natural Gas Company, was responsible for releasing this added radioactivity into the environment, as well as for producing solid radioactive products that remained in the well hole, possibly to contaminate underground water supplies.

While Gasbuggy was only a single experiment, the A.E.C., in partnership with Austral Oil Company of Houston, subsequently began the first of what promised to be a long series of even larger nuclear explosions, on the order of two 100-kiloton shots each year, for a period of 10 years or more. The first explosion, known as Project Rulison, was a 40-kiloton shot, some 8,400 feet below ground at a site near Rifle, Colorado, on September 10th, 1969. This time the flaring operations, which area residents tried to have permanently enjoined, were to release radioactive gases in an area containing towns, farming areas, grazing lands, and watersheds—from which Denver gets half its water! United States Public Health Service measurements following flaring from Gasbuggy revealed radioactivity 10 times normal in vegetation downwind from its site. The effects of Rulison, and subsequent explosions in Project Plowshare (which is still in progress), are therefore of genuine concern.

The A.E.C. did not bother to inform state and federal health officials in Colorado that part of Project Rulison called for the flaring of massive quantities of radioactivity into the atmosphere (20). Indeed, past history hardly encourages confidence in A.E.C. thoroughness where matters of safety outside of their own installations are involved. When the A.E.C., in partnership with private industry, came to Colorado in the early 1950s during the uranium boom, it had no standards for ventilation of the uranium mines, resulting in a lung cancer rate 100 times greater among uranium miners than in the general population.

The Commission admits that it has no practical plan for removing the radioactivity that makes natural gas freed by explosions such as Gasbuggy and Rulison much too dangerous to be used as fuel. Instead, it plans to rely on dilution to reduce the level of radioactivity, simply by adding the radioactive gas slowly to the gas distribution systems supplying homes and industry. Thus the ultimate effect will be to spread an admittedly lower dose of radiation to more people, despite the warnings of biomedical scientists, such as the Lawrence Radiation Laboratory's Arthur Tamplin, that no justification exists for exposing anyone to any amount of radiation, no matter how small. Besides the oil and natural gas experiments, the A.E.C. has plans for nuclear excavation projects to dig harbors and canals. Project Schooner, a 35-kiloton nuclear bomb exploded in Nevada on December 8, 1968, as part of this excavation program, created some of the highest levels of fallout in the western states since the end of atmospheric testing. Some of these tests have actually been responsible for radioactive fallout crossing our borders into Canada and Mexico, in direct violation of the limited test ban treaty (21).

The final area of possible hazard from underground nuclear explosions is at present the least understood, seismic effects. These earth-shaking effects can be divided into three categories: seismic waves generated by the nuclear explosion itself, earth movements triggered at the test site, and earthquakes triggered at some distance from the site. These underground explosions are similar to earthquakes; both are underground disturbances that displace rock and generate waves which travel through the Earth. Most underground tests are detectable on seismographs because these waves differ slightly from those caused by natural earthquakes. This has enabled both the U.S. and Soviets to monitor each other's underground tests from afar. Even a 1-kiloton nuclear explosion measures slightly less than 5 on the earthquake magnitude scale, which is comparable to a medium-sized natural earthquake. At the April 1, 1969 meeting of the Seismological Society of America, the A.E.C. finally admitted publicly what had been known for years, that faultless underground tests often create movements in the Earth's crust in the immediate vicinity of the test itself. The very real possibility also exists that underground tests can trigger local earthquakes that might add to the force of the explosion itself, as well as set off damaging quakes far from the test site. Preliminary studies have indicated a general increase of seismic activity in the Nevada region immediately following large underground tests. The larger tests planned for the immediate future might very well induce such activity in distant areas because of large shock waves traveling through the Earth's interior. Little is known about what it might take to damage the Earth's crust severely. Preparation for large underground tests at Amchitka Island in the Aleutians raised the possibility that these explosions may create tidal waves capable of doing damage in Hawaii, the California coast, or Japan. Earthquakes of the magnitude of some planned underground tests have created such tidal waves in the past, severely damaging areas in Japan and Hawaii. The record Chilean earthquake of 1960, for example, of magnitude similar to some planned tests, created 35-foot waves in Hawaii, thousands of miles from the center of the earthquake, killing 61 people and creating extensive property damage.

Controlled Atomic Power from Nuclear Reactors

When the atomic bomb was being developed during World War II, two routes were followed, each leading to the production of a particular kind of bomb. In one, the fissionable U-235 isotope, comprising 0.7 percent of natural uranium, was physically separated from the heavier U-238 that makes up the rest of natural uranium. This U-235 isotope fissions (Fig. 4–13A), and can be used to make uranium bombs. The uranium-238, however, absorbs neutrons (Fig. 4–12B), eventually to become element 94, known as plutonium (Pu). The plu-

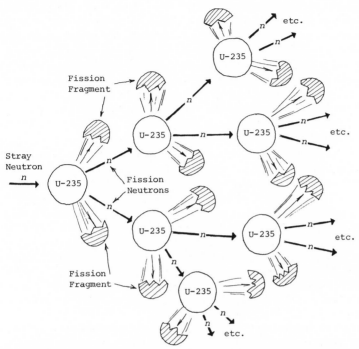

(A) Uncontrolled chain reaction occurring in atomic bomb of pure U-235.
Each fission produces several neutrons, causing a geometric increase
in the number of fissions produced in each succeeding generation:
1 fission in the first generation, 2 in the second, 4 in the third,
8 in the fourth, 16 in the fifth, and so on.

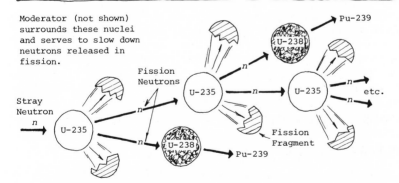

(B) Controlled chain reaction occurring in a nuclear reactor whose fuel
is a mixture of U-235 and U-238. The U-235 atoms fission, releasing
neutrons which are absorbed by U-238 and also cause other fissions.

Fig. 4–14. Comparison of uncontrolled (A) and controlled (B) nuclear fission.

tonium-239 isotope also fissions, and is therefore used to build plutonium bombs. In a second route, therefore, plutonium is made by enriching natural uranium with enough U-235 so that the neutrons released in the fission of U-235 nuclei can be absorbed by the U-238 nuclei, making them transmute into Pu-239. The neutrons from U-235 fissions are very fast moving, and must be slowed in order to be absorbed by U-238, as well as to be more effective in causing other U-235 nuclei to fission, providing still more neutrons. A material known as a *moderator* (graphite is often used) is placed around the enriched uranium "fuel" elements to make what is known as a nuclear reactor.

The difference between the uncontrolled fission in atom bombs and the kind of controlled fission in reactors is shown in Fig. 4—14. The bomb contains only U-235, so that all neutrons striking nuclei cause fissions which release more neutrons. Thus a stray neutron causes a fission, releasing 2 or more neutrons that go on to cause 2 or more fissions. Each of these, in turn, releases several neutrons, causing 4, then 8, then 16 or more fissions in succeeding generations (Fig. 4—14A). The tenth generation involves over a thousand fissions, the twentieth, over a million, all occurring in microseconds. In a reactor, however, only 1 neutron from each U-235 fission is allowed to cause other fissions (Fig. 4—14B). The other neutrons are absorbed by U-238 atoms, which retain them and change into plutonium (Fig. 4—12B). To ensure that this condition is maintained, control rods of boron or cadmium are inserted in the reactor, for these have the ability to absorb any extra neutrons that might tend to send the reaction out of control.

The reactor-fuel cycle and the production of radioactive by-products can be divided into several parts. First, uranium ore must be mined, milled, refined, purified, and made into fuel elements. There is little radioactive release at this stage, beyond the mines and processing plants. The second part of the cycle involves production of radioactive wastes formed during the reactor's operation. This includes fission products, normally remaining in the fuel, and radioactivity induced in the liquid or gas, often water or air, that must be used to cool the tremendous quantities of heat generated during reactor operation. Both the fuel elements and coolants are therefore potential sources of radioactive pollution caused by release of radioactive gaseous fission products such as iodine, xenon, and krypton into the atmosphere, by induced activity in atmospheric argon, or by release of other fission products such as strontium-90, cesium-137, barium-140, and zirconium-95. Normal operation of all reactors releases small amounts of radiation into the atmosphere. Research reactors at Brookhaven National Laboratory on Long Island and Oak Ridge National Laboratory in Tennessee, for example, are both air cooled, and emit large quantities of argon-41 produced by neutron capture of the argon-40 normally present in the cooling air (18).

The chemical reprocessing plants to which the irradiated spent reactor fuel is taken for disposal, however, present the most important potential source of radioactive atmospheric pollution. In these plants the fuel must be dissolved and processed to separate the unfissioned uranium and plutonium from the radioactive waste products. During the processing, fission-product gases such as krypton-85, xenon-133, and iodine-131 are released into the atmosphere. The amounts of these pollutants will increase as more chemical plants are built to handle the waste products of ever-increasing numbers of reactors. But the most drastic atmospheric radioactive contamination is not the result of normal re-processing but of accidents, some 71 of which were listed by the International Atomic Energy Agency as having occurred from 1945 to 1961. Six persons died of radioactive exposure in four of these incidents, but the most significant contamination of the surrounding environment occurred at Windscale, England, in 1957, when high levels of radioactivity spread downwind of the reactor site in a coastal strip 30 miles long and 6 to 10 miles wide. Airborne radioactive iodine deposited on the foliage in this area was consumed by dairy cattle, whose milk had such a high iodine content it was withheld from public consumption for several weeks.

Reactors will be considered in greater detail in Chap. 5, dealing with power sources, but an increase in reactor use must mean an increase in the radioactive fission products dispersed into the environment during routine operations and as a result of accidents. While little is known about what dose of radiation will produce what long-term effects in man, scientists generally agree that radiation exposures, no matter how small, carry risk of harmful biological effects.

It must finally be noted that the Atomic Energy Commission has accumulated some 75 million gallons of intensely radioactive wastes, mostly by-products of the reactor production of plutonium for nuclear weapons. Most of this is now stored in burial tanks at Hanford, Washington, in an area known for some previous earthquake activity. The possibility of additional earthquakes, or other accidents which might cause release of some of this radioactivity, is frightening indeed. The total amount of radioactivity now contained in these burial grounds has been estimated to be as much as would be released in a nuclear war (22).

Questions

1. Summarize, using chemical equations, the industrial production of nitrogen- and phosphorus-containing fertilizers. What are the pollution hazards of producing and using these chemical fertilizers?
2. What benefits to mankind did the introduction and widespread use of DDT bring after World War II? In view of its now known harmful effects, why is DDT still widely used in many parts of the world?
3. Imagine that you must decide whether or not DDT may be used. In attempting to reach and defend your decision, interview a farmer who has used insecticides, a chemical manufacturer of them, a conservationist. Attempt to answer their arguments. Conduct similar research for chemical fertilizers.
4. What insecticides are used in your local or regional area? What fertilizers? Have efforts been made to outlaw any? Who has supported these efforts? Who has opposed them?
5. How do organophosphate insecticides attack insects? How do they differ in their pollution effects from the chlorinated hydrocarbons?
6. Which radioactive atomic bomb fission products are especially dangerous to human health? Why are these particular isotopes so dangerous?
7. List the potential environmental hazards associated with underground nuclear explosions.

8. Briefly, indicate the difference between nuclear reactions in an atomic bomb, a so-called hydrogen bomb, a nuclear reactor. What are the potential environmental hazards associated with the operation of nuclear reactors?

Suggested Reading

American Association for the Advancement of Science, N.C. Brady, Editor, *Agriculture and the Quality of Our Environment.* A.A.A.S. Publication No. 85, Washington, D.C., 1967.

Collection of papers presented at the 133rd meeting of the A.A.A.S. in December 1966. Papers are grouped by the effects of agriculture on air, water, and soil quality. They are well written, largely nontechnical. A valuable reference.

Rachel Carson, *Silent Spring.* Fawcett Publications, Inc., Greenwich, Conn., 1962. A Fawcett Crest Book.

The classic, powerfully written, well-documented warning about the environmental effects of chemical pesticides such as DDT. Winner of 8 awards, this book is a devastating attack on human carelessness.

Herman F. Kraybill, Editor, *Biological Effects of Pesticides in Mammalian Systems.* Annals of the New York Academy of Sciences, Vol. 160, Art. 1, 1969.

A collection of highly technical papers, tracing in detail the chemistry and biochemical effects of pesticides. A valuable reference, but useful only to those familiar with organic chemistry and biochemistry.

Sheldon Novick, *The Careless Atom.* Dell Publishing Co., Inc., New York, 1969. A Delta Book.

Former editor of *Environment Magazine*, the author has presented for lay readers a history of the role of the A.E.C. in atomic energy development, with emphasis on the dangers of radiation and nuclear accidents.

Literature Cited

1. Paul R. Ehrlich, and Anne H. Ehrlich, *Population, Resources, Environment,* W. H. Freeman & Co., San Francisco, Calif., 1970.
2. Christopher J. Pratt, "Chemical Fertilizers," *Scientific American,* Vol. 212, No. 6, June 1965, pp. 62–72.
3. Howard R. Lewis, *With Every Breath You Take,* Crown Publishers, Inc., New York, N.Y., 1965, pp. 122–134.
4. George Kennedy, "Pollution Tarnishes the Golden State," *Philadelphia Inquirer,* May 14, 1970.
5. Francis J. Weiss, "Chemical Agriculture," *Scientific American,* Vol. 187, No. 2, August 1952, pp. 15–19.
6. Science and the Citizen, "DDT Shortage," *Scientific American,* Vol. 184, No. 4, April 1951, p. 32.
7. Jane E. Brody, "Use of DDT at a 20-Year Low," *The New York Times,* July 20, 1970, p. 16.

8. Barry Commoner, *Science and Survival*, Viking Press, New York, N.Y., 1967, pp. 22–23.
9. Robert L. Metcalf, "Insects vs. Insecticides," *Scientific American*, Vol. 187, No. 4, October 1952, pp. 21–25.
10. Rachel Carson, *Silent Spring*, Houghton-Mifflin Co., New York, N.Y., 1962; also in paperback: Fawcett-Crest, Fawcett Publications, Greenwich, Conn.
11. Hal Higdon, "Obituary for DDT (in Michigan)," *The New York Times Magazine*, July 6, 1969.
12. George M. Woodward, "Toxic Substances and Ecological Cycles," *Scientific American*, Vol. 216, No. 3, March 1967, pp. 24–31.
13. David R. Zimmerman, "Death Comes to the Peregrine Falcon," *The New York Times Magazine*, August 9, 1970.
14. Eugenia Keller, "The DDT Story," *Chemistry*, Vol. 43, No. 2, February 1970, pp. 8–12.
15. Research Reporter, "Pesticides and Our Environment," *Chemistry*, Vol. 39, No. 6, June 1966, pp. 25–26.
16. John Noble Wilford, "Deaths from Parathion, DDT Successor, Stir Concern," *The New York Times*, August 21, 1970, p. 1.
17. Donald G. Crosby, "Nonmetabolic Decomposition of Pesticides," in *Biological Effects of Pesticides in Mammalian Systems*, H.F. Kraybill, Ed., Annals of the New York Academy of Sciences, Vol. 160, Art. 1, June 23, 1969, pp. 82–96.
18. Air Conservation Commission, *Air Conservation*, Publication No. 80, American Association for the Advancement of Science, Washington, D.C., 1965, pp. 149–194.
19. Scientific Division, Committee for Environmental Information, "Underground Nuclear Testing," *Environment Magazine*, Vol. 11, No. 6, July–Aug. 1969.
20. Peter Metzger, "Project Gasbuggy and Catch-85," *The New York Times Magazine*, February 22, 1970.
21. Committee for Environmental Information, "Testing the Treaty," *Environment Magazine*, Vol. 11, No. 3, April 1969, pp. 10–11.
22. Sheldon Novick, "Earthquake at Giza," *Environment Magazine*, Vol. 12, No. 1, January–February, 1970.

5

Future Use of Land:
Can Increased Awareness
Bring Improvement?

Introduction

The study of the history of man's use of the land is not likely to fill one with much optimism about the long-term wisdom of our species. From neolithic beginnings, *Homo sapiens* has exploited the Earth for immediate benefit, first denuding it of its great stabilizing forests, then stripping it of nutrients by continued planting and taking—never returning. Natural balance has been upset by man's agriculture, and to make matters worse, he then added to ecological woes by introducing vast amounts of chemical fertilizers and synthetic pesticides with no thought about long-term effects of these substances. We studied these effects in the first part of Chap. 4, and know that they are widespread and serious. Just how serious, in the long run, is not yet fully understood.

Our mushrooming population, with its ever-increasing standard of living giving rise to greater and greater needs, has built and moved in large numbers to great cities. Here, in a completely artificial environment, where even climate is different from that of the neighboring countryside, people are crowded together, beset by noise and pollutants from the great numbers of automobiles as well as heavy concentrations of industry. It is no accident that past air pollution disasters occurred in cities or heavily industrialized valleys. The problems of disposing of the trash and of supplying electric power for ever increasing populations will have to be faced.

Perhaps the most serious immediate crisis, because it contributes so heavily to our present air pollution and trash disposal problems, is the transportation dilemma: how we get about from one place to another. Thus far, this has been very largely by gasoline-powered vehicles, even though they emit more than half of all air pollutants, while abandoned cars increasingly litter our streets. It is to this problem, therefore, that we turn first.

The Internal-Combustion Engine Automobile

Internal-combustion engines power automobiles, trucks, buses, aircraft, and marine vessels, but the automobile is the dominant source of air pollution. Although transportation accounts for only 20 percent of total energy use in the

United States, it results in 60 percent of all air pollution nationwide, and as much as 90 percent in those urban areas where air pollution controls on stationary sources are enforced (1). The actual atmospheric pollutants from auto engines come from several sources. Exhaust emissions from the tailpipe, according to the President's Environmental Pollution Panel (2), account for some two-thirds of hydrocarbon emissions from autos (see Table 5–1). Other emission points are crankcase venting and carburetor and gasoline tank evaporation.

Why unburned hydrocarbons should escape as emissions has to do with combustion conditions in the engine itself. If we assume gasoline has the single formula $C_8 H_{18}$, the equation for its complete combustion would be:

$$C_8 H_{18} + 12\tfrac{1}{2} O_2 \longrightarrow 8CO_2 + 9H_2 O$$

Since air is some one-fifth oxygen, 62.5 moles of air would be required for such complete combustion. Comparing molecular weights (114 for gas, about 29 for air) gives a weight ratio of 114 gas to 62.5 × 29, or 1800, for air, about 1 to 17. The average molecular weights for actual gasoline mixtures are about 124, giving gasoline-to-air weight ratios of 1 to 14.5 for complete so-called stoichiometric mixtures. If less air is used, the mixture is said to be rich (in gasoline), and produces deadly carbon monoxide (CO) and unburned hydrocarbons during the incomplete burning that results. If the weight of air is more than 15 times that of the gasoline, the mixture is said to be lean, and produces less carbon monoxide and unburned hydrocarbons. Indeed, studies show that carbon monoxide levels of 15 percent by volume of exhaust gas composition, at a 9 to 1 air-gas weight mixture, can be reduced almost to zero around the 14.5 to 1

Table 5-1. Contribution of the Various Engine Modes to the Total of Exhaust Hydrocarbon Emission, Excluding Warm-up*

Mode	Percent of total† contributed by each mode (± Standard Error)	
	Vehicles without emission control	Vehicles with emission control
Idle	5 ± 1	3 ± 1
0–25 mph Accelerate	20 ± 3	24 ± 6
30 mph Cruise	7 ± 2	9 ± 2
30–15 mph Decelerate	13 ± 5	8 ± 6
15 mph Cruise	4 ± 1	3 ± 1
15–30 mph Accelerate	32 ± 5	35 ± 8
50–20 mph Decelerate	19 ± 6	17 ± 8
Number of vehicles tested	50	76

*Data contributed by General Motors Engineering Staff, Michigan. From Hurn (2).
†Composite of exhaust emissions produced during the seven modes of a hot cycle as prescribed by U.S. federal standards for certification of 1968 vehicles.

stoichiometric point (2). The problem with lean mixtures, however, is that the engine misfires, resulting in rising emissions of other categories.

Since nitrogen-oxygen combination—first to produce nitric oxide (NO) with subsequent further oxidation to the dioxide (NO_2)—is favored by high combustion temperatures as well as the availability of both oxygen and nitrogen (i.e., excess air which contains free, unused oxygen), mixtures a bit on the lean side of the 1 to 14.5 stoichiometric ratio produce the highest nitrogen oxide concentrations. Exhaust emissions also produce large numbers of extremely fine particles, both organic and inorganic in nature, amounting to an average of 0.78 mg per gram of gasoline burned (2). The most significant part of this is the lead compounds resulting from the use of tetraethyl lead [$Pb(C_2H_5)_4$] as a fuel additive to provide antiknock characteristics necessary for most present-day high-compression engines. The lead compounds are mostly in the form of mixtures of $PbCl \cdot Br$ and $NH_4Cl \cdot 2PbCl \cdot Br$ (2).

Evaporative emissions, from both fuel tank and carburetor, account for from 10 to 30 percent of the hydrocarbons in auto emissions, with losses from the carburetor (Fig. 5-1) occurring primarily during periods after a fully warmed-up engine is stopped. This is because the cooling underhood air flow stops when the engine stops, so that the carburetor absorbs heat from the hot engine components, heating gasoline in the carburetor as much as $50°$ to $80°F$ above normal, which is well into the boiling range of any gasoline (2). Emissions from unburned gas blown past the piston rings into the crankcase during the combustion cycle have been eliminated in U.S. cars beginning with 1963 models, by recyling this blown-by gas back into the engine intake instead of venting it into the atmosphere. Fuel evaporation from the gas tank, carburetor, or elsewhere in the fuel system can be minimized, or eliminated, says the automobile industry, by leading such vapors through a complex system into storage areas, such as the engine crankcase or a canister of activated charcoal granules. Presumably, the gasoline vapors would be stored in such areas until the engine is restarted, at which time they would be routed back to the combustion chamber and burned (4). The main problem with such storage systems is that carburetion is upset when the vapors are introduced to the engine. An alternate method for reducing evaporative emissions would be to change the properties of gasoline, something that would have to be accompanied by changes in engine design. It could theoretically be done either by using longer-chain hydrocarbons to reduce volatility, or by replacing four- and five-carbon double-bonded hydrocarbons with less reactive similar-sized saturated hydrocarbons. While the latter change would not reduce hydrocarbon emissions, it would reduce the reactive hydrocarbon emissions.

Exhaust emission controls tried or proposed include the fuel modification described above, devices that treat the exhaust gases, and devices that modify engine operating conditions. This latter approach has been most widely used to date, and involves adjusting such things as fuel-air ratio, spark timing, and cylinder design, but the first two need constant adjustment, and, as we have already seen, reducing hydrocarbon and carbon monoxide emissions results in increased emissions of nitrogen oxides. Afterburners and catalytic converters (Fig. 5-2) show the most promise as control devices for exhaust gases. In the

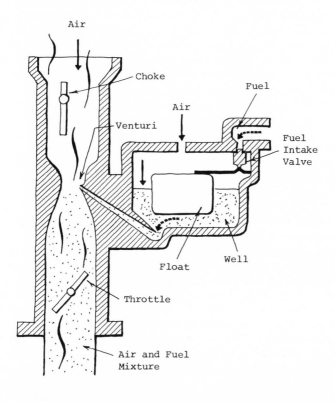

Air

Choke

Fuel

Air

Venturi

Fuel
Intake
Valve

Well

Float

Throttle

Air and Fuel
Mixture

Fig. 5–1. Detail of carburetor in gasoline engine. Air rushing through the narrow "venturi" tube creates a low pressure, so that air above the well, at normal pressure, forces gasoline to spray into the air stream, forming an explosive mixture which goes to the cylinders of the engine. Turning the throttle controls the amount of mixture that gets into the cylinders and thus controls the speed of the engine. For cold weather starting, the choke is turned to limit the air supply and let more gasoline into the cylinders. After Ruchlis and Lemon (3).

afterburner, air is admitted to the exhaust manifold, and a direct flame provides some combustion of unburned gases. Again, while reducing hydrocarbon and carbon monoxide emissions, it may increase nitrogen oxide emissions. The catalytic converter is designed to convert unburned hydrocarbons and carbon monoxide to water and carbon dioxide by further combustion. In theory, it could also reduce the nitrogen oxide emissions by allowing carbon monoxide and nitric oxide to react, forming nitrogen and carbon dioxide:

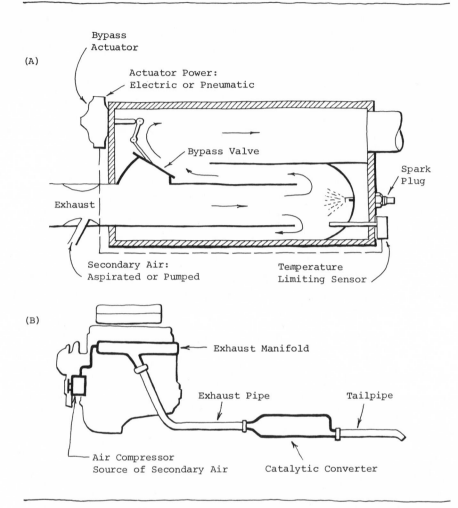

Fig. 5–2. Two methods of exhaust emissions control for automobile engines:
(A) Direct-flame afterburner. (B) Catalytic converter system. After Hurn (2).

$$CO + NO = \tfrac{1}{2}N_2 + CO_2$$

This reaction, unfortunately, requires a reducing atmosphere, as opposed to the oxidizing atmosphere required to treat hydrocarbons and any additional carbon monoxide. Another disadvantage of the catalytic converter is its gradual decrease of effectiveness with use, due to fouling by lead compounds, as well as deactivation of the catalyst itself by high temperature.

The only other suggested method of nitrogen oxide control is to reintroduce a portion of the exhaust gas into the cylinder intake, which serves both to reduce peak combustion temperature and reduce absolute oxygen concentration, which together can reduce nitrogen oxide emissions as much as 90 percent (2). To achieve such reductions, however, up to 30 percent exhaust gas must be used in the intake charge, which drastically decreases the maximum power available from the engine. Thus the difficulties of reducing emissions of nitrogen oxides as well as those of carbon monoxide and hydrocarbons are serious indeed. Even with effective control methods (they do not seem likely), continued mainte-nance would be an awesome problem. Installation on the millions of cars already being driven could prove prohibitive in cost, while the ever-increasing number of cars on the road would make total pollutant emissions not much less than now, even with the controls we have mentioned.

Despite the clear evidence that automobiles are responsible for a substan-tial amount of pollution, especially in cities and industrial areas, more and more Americans are driving ever more cars, so that even if the emission controls previously mentioned cut the amount of pollution per car, the total amount of pollutants emitted may well continue to increase with continued use of the internal-combustion gasoline engine. Some 100 million autos were on our roads at the close of the 1960s, with Detroit supplying new ones at the rate of 22,000 per day. Auto industry spokesmen claim this rate can be doubled by 1980, so that some 170 million autos might be expected on our roads by 1985 (5). To reduce this source of pollution, therefore, either the number of autos must be drastically reduced, or their design drastically changed. A number of possibilities do exist, and will be considered here.

The Steam Car

The use of steam as a source of motive power (Fig. 5–3) in locomotives dates from about 1830, while the Stanley twins produced their famous "Steam-er" cars at the beginning of the twentieth century. Their cars were fast and safe, but the twins refused to mass-produce them, being uninterested in the business end of things. In 1906, however, a Stanley "Rocket" steam car set five world speed records at Daytona Beach, Florida, and was once clocked at better than 127 miles per hour! (6). Just prior to selling their business in 1917, the Stanleys produced some 600 to 700 cars annually.

Because of the pollution problem, interest has been revived in the steam car. Its power unit would consist of a burner, a boiler, the actual engine, and a condenser. The burner heats water, or some other liquid, in the boiler, producing steam. Burner fuel could be kerosene, or almost any other flammable substance, and it is burned externally. This means it could be burned evenly and com-pletely, instead of exploded and burned incompletely, as in the internal-combustion gas engine. As a result, emissions would be greatly reduced. Tests run on an 8-year-old Williams steam car, for example, show it releases 1/450th the unburned hydrocarbons, 1/70th the carbon monoxide, and 1/40th the nitrogen oxides as a modern, uncontrolled internal combustion engine (7).

To move the car, the engine valve is opened, letting steam into the engine block, where it performs a function similar to the air-gas mixture in an internal-combustion engine. When steam enters the cylinders and pushes the pistons forward, this motion is transmitted to a crankshaft (Fig. 5–3), which transfers

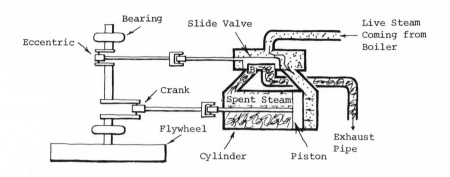

Fig. 5–3. Principle of the steam engine. Steam from the boiler enters through port A and pushes the piston to the left, while spent steam from the previous cycle goes out the exhaust through port B. The moving piston pushes the rod, causing the crank to turn the heavy flywheel. The off-center eccentric on the shaft of the flywheel is thus rotated and pushes the slide valve over to the right to close port A and open port B. Steam now enters port B and pushes the piston back toward the right, while spent steam goes out the exhaust through port A. As the flywheel continues to turn, the eccentric also rotates, pulling the slide valve back to its original position so that steam can enter through port A. The repetition of the whole process keeps the engine going. After Ruchlis and Lemon (3).

power to the wheels. Instead of being exhausted, as shown in the illustration, the spent steam can be recondensed to water, and recirculated, to be used over and over again. The so-called Rankine engine is believed to be the most practical version available at the present time, according to a Senate Commerce Committee study (7), and is being used at present to power California Highway Patrol Dodge Polaras. It can deliver a working fluid, transformed to a vapor, either to drive pistons, as described above, or to drive a turbine. The turbine consists of a bladed wheel which can be driven by jets of steam (Fig. 5–4A) or hot gases (Fig. 5–4B). There is little doubt that, given the money and development effort that has gone into gasoline engines, steam cars could be quickly developed.

The Electric Car

Like the Stanley Steamer of the early 1900s, electric cars were popular at the beginning of this century, but had vanished by the late 1920s to be replaced by the noisier, more polluting, but more powerful gasoline engined car. Some people feel that with proper research and development the electric car can, eventually, replace the present internal-combustion-engine-powered auto. While it is true that electricity to recharge the batteries necessary to run any future car's electric motor must be made by burning potentially polluting fuel, this would be done at stationary plants, whose emissions could be more effectively

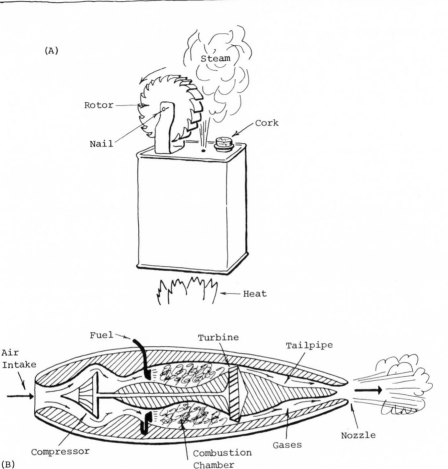

Fig. 5–4. (A) Principle of the steam turbine. A rod through the center of the turbine wheel might operate a generator, to produce electricity, or might drive a system of gears, to provide power to the wheels of a car. (B) Gas turbine, as used in a jet engine. Compressor pushes air into the combustion chamber, where it is mixed with fuel and burned. The hot expanding gases rush out the back, providing the forward thrust of the engine, and simultaneously turning the turbine, which operates the compressor at the front. After Ruchlis and Lemon (3).

controlled than can the emissions from moving autos burning gasoline. Also, supplies of crude oil, and hence the gasoline made from them, would be used up less quickly if the need for the hundreds of millions of gallons of gasoline to power the hundreds of millions of present auto engines were to cease.

While the batteries and motors in the front and rear of an electric car can be expected to weigh much more than the engine and transmission of a gasoline-powered auto, this is more than compensated for by elimination of many weighty items in present cars that are not needed in a vehicle run by electricity. An electric car needs no transmission hump in the middle of the floor, and it requires little of the insulation now required to shield an auto's interior from the noise, heat, and fumes of the engine. Furthermore, it needs no radiator, no grill, no fuel tank, no fuel pump and plumbing, no exhaust pipes and mufflers. Using already existing electric motors, an electric car's overall performance compares favorably with present high-compression internal-combustion autos. This was demonstrated in 1964 when an automobile company's engineers took a compact model off its production line, replaced its gasoline engine with an electric motor, and clocked the car at a top speed of more than 80 miles per hour (8).

Electric motors offer obvious advantages over internal-combustion engines. They are much simpler in design, need almost no maintenance, can be used in various arrays. It is the battery problem that has been the principal obstacle to practical electric automobiles, for the kind of battery that delivers electricity to run a motor determines the range of the car. This is because the distance traveled between battery recharges (the "refueling") obviously depends on how much electrical energy can be packed into such a battery. The common so-called "dry cell" battery operates on the same principal as the chemists' voltaic "wet" cell, except that it uses a paste in place of an actual solution. The principal of operation of a typical voltaic cell, the Daniell cell, is shown in Fig. 5—5. It depends for its operation, as do all chemical cells, on the ability of some substances to react spontaneously with others by transferring electrons to them. Thus, if the two substances are suitably separated, such electron transfer must take place through an external wire, and hence produces an electric current.

Both the dry cell battery and the Daniell cell described in Fig. 5—5 are examples of primary cells. This means that they are irreversible; once the initial chemical (zinc anode and copper ions in the Daniell cell) reactants are used up, the cell is dead and useless—it cannot be recharged and reused. The lead storage cell, or battery, so widely used in automobiles today, is an example of a secondary, or reversible cell (Fig. 5—6). When new, its different electrodes (spongy lead anode and lead dioxide cathode) react spontaneously with each other and their sulfuric acid solution electrolyte, producing identical lead sulfate electrodes. As this happens, however, the car's generator transforms the energy of motion of the wheels into enough electrical energy to force the electron flow to reverse. This forces the previous anode and cathode reactions to reverse themselves, producing the original lead and lead dioxide electrodes, thus recharging the battery and enabling it once again to produce electricity.

It is one thing to use the above-mentioned lead storage cell to start an internal-combustion engine and fire spark plugs, using little electricity and being constantly recharged. It is quite another order to supply electricity continuously

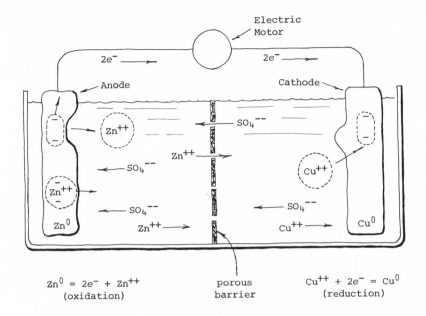

$Zn^0 = 2e^- + Zn^{++}$ porous $Cu^{++} + 2e^- = Cu^0$
(oxidation)   barrier   (reduction)

overall: $Zn^0 + Cu^{++} = Zn^{++} + Cu^0$ + electricity

Since zinc (Zn^0) has a greater tendency to lose electrons than does copper (Cu^0), the spontaneous chemical reaction·in which Zn loses electrons to Cu^{++} ions produces an electric current. This is because the zinc anode is physically separated from the Cu^{++} ions, so the electrons (e^-) can be transferred only through an external wire.

First, some Zn^0 atoms on the zinc anode go into solution as Zn^{++} ions, leaving 2 electrons per atom behind on the anode. These electrons repel each other through the wire (and electric motor), eventually arriving at the copper cathode. There they attract Cu^{++} ions from the solution, with which they combine to form a plating of copper atoms on the cathode.

The two "halves" of this so-called cell are separated by a porous barrier through which ions can pass. Each time a Zn^{++} ion goes into solution from the anode, either a Zn^{++} ion already in solution must pass out (to the right) of the anode side, or an SO_4^{--} ion must pass in (from right to left) to this side, to keep electrical neutrality. Each Cu^{++} that deposits onto the cathode, leaving the solution, must be accompanied by either an SO_4^{--} going through barrier to left side or a Zn^{++} coming through barrier to right side.

The cell is dead when zinc anode completely dissolves, or all Cu^{++} ions have been plated out on the cathode.

Fig. 5—5. Typical voltaic cell: the Daniell cell.

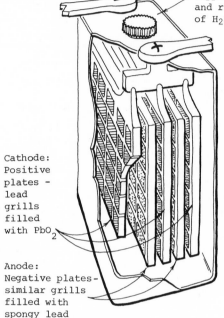

Capped hole for testing
and replenishing electrolyte
of H_2SO_4 and distilled water

Discharge: Spontaneous chemical
reactions produce
electricity.

At Lead Anode:

$$Pb^0 = 2e^- + Pb^{++}$$

(the Pb^{++} produced combines with
SO_4^{--} ions in the H_2SO_4 electrolyte to give a white precipitate
of $PbSO_4$)

At Lead Dioxide Cathode:

$$2e^- + PbO_2 + 4H^+ + SO_4^{--} = PbSO_4(s)$$

(the H^+ and SO_4^{--} are furnished
by the H_2SO_4 electrolyte)

The cell is dead when both electrodes have been converted to
$PbSO_4(s)$.

Cathode:
Positive
plates -
lead
grills
filled
with PbO_2

Anode:
Negative plates-
similar grills
filled with
spongy lead

Lead storage cell can be recharged by supplying electricity to
both sides, forcing the $PbSO_4(s)$ to react:

Charge: Electricity (from generator) forces non-
spontaneous chemical reactions.

At (former) lead anode (now $PbSO_4$):

$$2e^- + PbSO_4 = Pb^0 + SO_4^{--}$$

At (former) lead dioxide cathode (now $PbSO_4$):

$$2H_2O + PbSO_4 = PbO_2 + 4H^+ + SO_4^{--} + 2e^-$$

The cell is thus restored to its former condition, ready to
discharge electricity again.

Fig. 5–6. Lead storage cell: a secondary, reversible cell. Illustration modified
from Linus Pauling, *College Chemistry*, W.H. Freeman & Co., San Francisco,
Calif., 1950, p. 134.

to run an electric motor which has no other source of power, like the gasoline that runs the internal-combustion engine. If the lead storage cell were used to power an electric motor car, such a car would have a range of less than 20 miles, in city driving, before it needed recharging. Even the very costly silver-zinc batteries developed for the electrical systems of space vehicles would give an electric car a range of only 50 miles in city driving (8). The most promising device for electric cars, still in the experimental stage, is known as the air battery. The fuel is a metal (zinc is a current favorite), which is oxidized by (loses electrons to) oxygen in pressurized air that is pumped to the metal electrodes. The electric current is generated by the reaction between the metal and oxygen to form the metallic oxide, which is stored in a separate bank for spent fuel (8).

To recharge such a battery, an outside voltage is applied across its terminals, forcing the stored metallic oxide to decompose back into the metal, which is electroplated onto the anode up to a desired thickness, and oxygen, which is vented to the atmosphere. Now the battery is fully charged again, ready to repeat its energy-delivering cycle. Experiments with such batteries, using zinc as the metal, indicate ranges of 80 to 160 miles for an electric car, depending upon driving conditions (8). The design of such an air battery is shown in Fig. 5–7. Its refueling need present no problem, for it could be designed in the form of

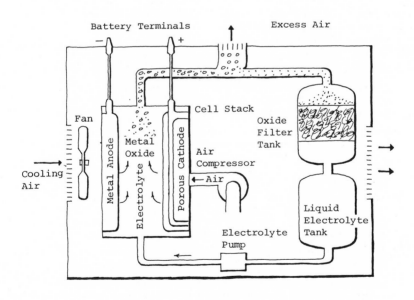

Fig. 5–7. The metal-air battery. After Hoffman (8).

standardized packs that could be quickly exchanged at "gas" stations, or recharged more leisurely in the home garage with simple equipment at night (8). Thus, while a modern electric car will not be produced quickly, the evidence indicates that further research and development of such a car is a worthwhile investment.

The Flywheel-Powered Automobile

The most novel possible replacement for the internal-combustion engine automobile is a car or bus powered either by a rapidly spinning flywheel alone, or by a flywheel coupled to another engine. While the type of flywheel needed to store the amounts of energy required to power vehicles would have to be based on complicated, not yet fully developed design considerations, the principle of any flywheel is the same as that which keeps a child's top spinning. The motor vehicle flywheel would consist basically of a disc or streamlined bar flywheel (Fig. 5–8) that could be spun to high speed by an electric power motor plugged into a special outlet in a charge. Some kind of power transmission would then be used to couple the flywheel to the wheels of the vehicle to propel it (9).

There are several different types of flywheels (Fig. 5–8), and the energy-storage potential of each is determined by the complex inter-relationship of its mass, shape, cross-section, density, composition, etc. (9):

1. *Weighted-rim flywheel*—an early type, widely used today in balancing automobile engines and heavy drill operations.
2. *Plain-disc flywheel*—usually of solid steel, a shape many might not think of as a flywheel; it stores half again as much energy as the weighted-rim type.
3. *Tapered-disc flywheel*—the most effective shape for a conventional flywheel, it has twice the energy-storage potential of the weighted rim.
4. *Streamlined-bar flywheel*—composed of music wire strands stacked into a bar and held together in the center by a yoke containing drive and support shafts; it can store four times the amount of energy as the weighted-rim.
5. *Super flywheel*—improves on the streamlined bar in the same way that the tapered disc is superior to the plain-disc type. Like the streamlined bar, it is made up of music wire and other filaments bound by resins; it stores more than six times the energy potential of the weighted-rim type.

The super flywheel seems to be the most practical design for flywheel power currently in use. Further design improvements would be expected to increase efficiency by reducing the drag of air resistance as the flywheel is rotated.

Land vehicles using flywheel power date back to 1904, although the success of the concept was not convincingly proved until 1953 with the introduction, in Switzerland, of the Oerlikon bus, a 70-passenger vehicle that served successfully for many years in Europe and Africa. The bus glided smoothly for more than half a mile under the sole power of a 3,300-pound flywheel, which was rewound electrically at each bus stop while passengers boarded and left the vehicle (9). It took 2 minutes to get the flywheel spinning at high speed so that the bus could cruise away with its flywheel driving the electric-charging motor in reverse, converting it to a generator. This generator powered other electric

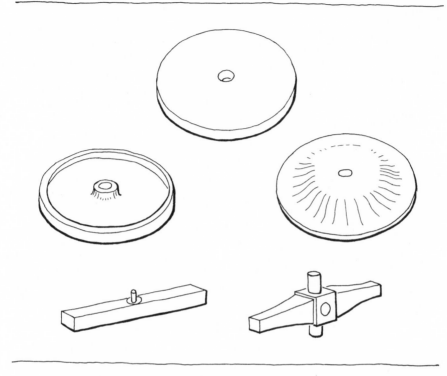

Fig. 5–8. Flywheel shapes. After Hohenemser and McCaull (9).

motors, which in turn drove the wheels (9). With a super flywheel of the future, such a bus might be able to operate all day without recharging.

A smaller, lighter vehicle, weighing less than 1,300 pounds, with a super flywheel mounted horizontally in the rear, could travel about 100 miles before it needed recharging; it would have a maximum speed of 70 miles per hour, and an acceleration time of 15 seconds to go from 0 to 60 miles per hour, according to laboratory tests with super flywheels (9). Not only would fuel costs be virtually eliminated, but the only pollution would be from electrical power plants generating the recharging electricity. We have already mentioned that it is technically easier and more economical to control pollution emissions at stationary power plants than it is on internal-combustion-engine-powered automobiles.

Large cities such as New York could drastically cut their smog by replacing internal-combustion-engine-powered buses with flywheel-powered vehicles. Because large quantities of electric power are available at night in New York City, fleets of such buses could be recharged then, and, depending on flywheel design, possibly store enough energy to power them for the entire next day. Thus New York City could virtually eliminate its present fleet of 4,300 diesel-powered, smog- and odor-producing buses.

Solutions to Automobile Pollution

There thus appear to be some potentially viable alternatives to the present pollution-emitting internal-combustion-engine-powered automobiles. Besides the few examples discussed here, there are many other possibilities, including some rather interesting compromise combinations. One such compromise would be an electric car with a small gasoline-engine-driven generator to recharge its batteries (10). While this engine would burn gasoline, it would be much smaller than present auto engines, and designed to operate at constant speed and load, which would simplify its design and minimize its pollutant emissions.

While a final answer is not yet at hand, there is every reason to be optimistic about eventually finding one. The present automobile was developed only with continued expenditure of time, resources, and a great deal of money. We cannot expect fo find a suitable replacement for it for nothing, but surely the elimination of more than half our present air pollution problems should be worth some effort.

Energy Requirements and Supplies

There is an obvious connection between population growth, economic development, and energy requirements. Most available energy comes, ultimately, from mineral sources: coal, oil, natural gas, and uranium-containing minerals. These minerals are used as fuel for the heat energy which keeps us warm in cold climates and also produces most of our electricity. The industrial age has so accelerated the demand for these as well as other minerals that man managed to mine and consume more of this planet's mineral resources in the period between 1918 and 1940 than in all preceding history. And it is reliably estimated (11) that world consumption of mineral fuels in the last 40 years of this twentieth century will equal about three times the amount consumed from the dawn of history through 1960!

Every science text details how Michael Faraday's discoveries in the first half of the nineteenth century led to the development of modern electric generators. These texts point out that electricity is produced when a coil of wire moves through a magnetic field (Fig. 5–9), and that such motion usually requires heat energy to produce the steam needed to drive motion-supplying turbines (Fig. 5–4). Only when the motion of falling water can be used to drive water wheels is the necessity of producing heat avoided. Most of the needed heat energy is produced by burning fossil fuels such as coal, or, increasingly these days, by using enriched uranium as nuclear fuel, and utilizing the heat released during nuclear fission of the uranium-235 isotope present in the fuel.

Whether the fuels be fossil (coal, oil, natural gas) or nuclear (uranium), they are vanishing assets. Once used, they are gone forever. Known coal supplies could last for several thousand years. However, the significant question about coal, according to a National Academy of Sciences report (12), is not how long it will last, but rather over what period of time it can serve as a major source of industrial energy. If coal alone continued to be used as the main industrial energy source, the more readily available middle 80 percent of the world's coal resources would be exhausted in less than 200 years. Even if other fossil fuels

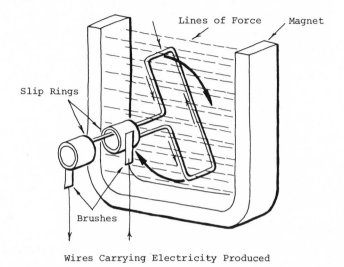

Lines of Force Magnet

Slip Rings

Brushes

Wires Carrying Electricity Produced

Fig. 5–9. A simple generator. According to Faraday's principle, current is produced when lines of force between magnetic poles are cut by a moving wire. The rotating coil is connected to slip rings which are in contact with brushes; the brushes conduct the electric current from the coil to wires, and then to a lamp or other device. Using a larger magnet or increasing the number of turns of wire on the rotating coil will increase the amount of electricity generated.

(oil and natural gas) shared the burden of supplying most of the world's energy requirements, that middle 80 percent of coal resources would not last beyond 300 or 400 years (12). The world's uranium reserves, if used as a fuel source, would not last more than 20 or 30 years!

Fossil Fuel and Nuclear Power Similarities

It is important to realize the differences between fossil fuel and nuclear electric power plants. First, however, let us consider the similarities common to both types. Both fuels are at present used almost entirely in steam-electric-generating plants—i.e., to heat water so that the resulting steam may turn steam turbines (Fig. 5–4). Thus, both types of fuel create similar problems of thermal pollution because both types are relatively inefficient. In each case, only about one-third of the heat produced by the fuel goes to produce electricity. The major portion of it is discharged into the environment as waste heat. Both types of fuel pollute water with chemical additives, and pollute air with sulfur and nitrogen oxides or radioactive materials. Both types of power plants at times use large amounts of land to transmit electric power from plant to consumer (13).

The principle by which electricity is produced in both fossil and nuclear steam-generating plants is straightforward, and of course the same regardless of the type of fuel used. The fuel heats water, forming steam, which is piped to a turbine, causing the turbine blades to turn (Fig. 5–4). The turbine turns a generator, which converts the mechanical energy of motion to electricity (Fig. 5–9). This electricity is then sent through transformers and switches, and ultimately reaches the consumer. After leaving the turbine, the spent steam goes to a condenser where every gram of it that condenses to water releases 540 calories, much of which is carried away by coolant water, giving rise to thermal pollution where the coolant water is discharged into lakes or rivers. While fossil plants discharge some of their waste heat into the air, essentially all waste heat from nuclear plants goes into our water supplies (13). As the amount of power generated increases, the size of the individual plants increases, and so does the amount of heat discharged to the environment.

Typical power plants now being built and planned will have capacities of between 500 million and 1,000 million watts of electrical energy—usually written as 500 to 1,000 megawatts, since 1 megawatt equals 1 million watts. A 100-watt electric light bulb, burning for 10 hours, uses 1,000 watt-hours, or 1 kilowatt-hour, or one-thousandth of a megawatt-hour of electricity. It has been

Table 5-2. Comparative Costs of Cooling Water Systems for
Steam-Electric Plants*

Type of system	Investment cost of 1,000-megawatt plant	
	Fossil fuel	Nuclear fuel
Once-through, no control measures, heat discharged directly into a fresh water river or lake	This method taken as baseline—it is the minimum cost option; no added costs	
Once-through, no control measures, heat discharged directly into a salt water source	$2 to 3 million	$3 to 5 million
Cooling ponds with recirculation of the water from the pond (approximately 1,000 acres needed for a plant of this size)	$4 to 6 million	$6 to 9 million
Wet cooling towers, most heat loss to atmosphere via evaporation with resultant production of fogs, mists, and considerable water loss	$5 to 9 million	$8 to 13 million
Dry cooling towers, heat transferred to atmosphere without evaporation, no resulting fog or mist. To date no such towers have been constructed for a large power station	perhaps $20 million	perhaps $25 million

*From Abrahamson (13).

estimated (13) that a 1,000-megawatt fossil fuel plant will discharge over 400,000 kilocalories of heat per second as waste heat, while present less-efficient nuclear plants of the same capacity will release over 600,000 kilocalories of heat per second. Such amounts of heat could, if properly utilized, supply all the winter heating needs of 300,000 homes located in the coldest parts of the United States (13)! If utilities continue to operate as they have in the past, however, all this heat will be released to the environment at a single point in a lake or river, with no control measures. This is the cheapest way for utilities, since environmental damage does not yet appear on annual balance sheets.

It has been found that the temperature rises caused by dumping waste heat into a body of water may kill or render more susceptible to disease certain aquatic organisms, as well as disrupt normal biological rhythms. Since dissolved gases such as oxygen are known to be less soluble at higher temperatures, oxygen concentrations are decreased just at the time the aquatic life needs more oxygen to function at the higher temperatures. In addition, the increase in rooted plant growth caused by the higher temperatures leads to decreased river flow rates and increased silting.

Waste heat must be removed from power plants if they are to continue to function, so it is obvious that some control of this waste heat is needed if the environment is to be spared. To date, little research has been conducted to find a practical use for this waste heat. Such use might include warming farmland or large greenhouses or combining electric plants with ocean-water desalting plants. The more immediate practical solutions are to use control devices, such as evaporative "wet" towers, in which the heated water mixes with air and is cooled largely by evaporation, or dry types of cooling towers, in which the water and air do not directly mix; instead, the heated water flows through a system of pipes and channels, and the cooling air is drawn across this system by fans. These controls cost money, however, not only because of the capital cost of the towers, but because of the continuing need for responsible treatment of the waste heat to avoid thermal pollution. Tables 5–2 and 5–3 detail such costs, and they

Table 5-3. Percent Cost Increase because of Thermal Pollution Control for the Average U.S. Consumer*

Type of cooling system	Class of consumer		
	Industrial	Commercial	Residential
Once-through—fresh water	No added costs; the base situation. All costs listed below are increases over this situation		
Once-through—seawater	0.34%	0.16%	0.14%
Cooling pond	0.94%	0.43%	0.39%
Wet mechanical draft tower	3.17%	1.41%	1.28%
Wet natural draft cooling tower	1.48%	0.68%	0.62%
Dry cooling tower	Very uncertain, may be in the range of 1.5% to over 3% for residential consumers		

*From Abrahamson (13).

are considerable. But it must be remembered that if some type of control measures are not undertaken, there will be considerable environmental costs, some of which will be immediately and practically felt. This is because increasing the temperature of a body of water will substantially reduce its capacity to assimilate waste. Alabama's Coosa River, for example, had its temperature raised only $9°F$ above the existing summer temperature of $77°F$, which resulted in a reduction of the waste assimilating capacity of the river by 11,000 pounds per day of oxygen-demanding wastes (13).

Besides the thermal pollution common to both types of electric power plants, both use and discharge chemical pollutants, such as chlorine used to protect the machinery, and biocides added to prevent buildup of slime in condensers. As of 1972, most of these additives were discharged into the nearest body of water, although it is possible to contain them. Finally, transmission lines, whether from fossil or nuclear plants, require some 100 acres of land per 1 mile of lines; thus, the farther away a plant is from consumers, the more land is used. Of course, plants close to population centers expose large numbers of people to their various pollutants. Close by or far away, the consumer loses something from the environment—land or clean air.

Unique Features of Fossil Fuel Plants

Except for motor vehicles, fossil power plants are the largest single source of sulfur and nitrogen oxide pollution, accounting for almost half of the total tonnage of SO_2, and between one-fourth and one-fifth of total nitrogen oxide tonnage in 1965 (14). Although low-sulfur coal deposits do exist, they are often far from major metropolitan areas, and are used largely in metallurgical operations. As a result, more than 90 percent of the coal burned in United States power plants through the 1960s contained more than 1 percent sulfur (14), thereby giving rise to SO_2 released to the atmosphere. Control measures for SO_2 do exist, and their employment, regardless of cost, is a necessity, if air quality is to be improved or even maintained.

Alternatives to SO_2-control measures have been adopted by many utilities faced with regulations prohibiting the use of coal containing more than 1 percent of sulfur. These have been to switch to natural gas and oil as fuel, in place of coal. Most fuel oil comes from abroad, and is high in sulfur content and hence must also be specially treated. On the other hand, available supplies of natural gas, mostly methane (CH_4), are inadequate to meet such demands, and have given rise to the fuel shortages so widely reported in the press beginning with the latter half of 1970. Yet another factor contributing to this shortage was the refusal of many utilities to renew long-term contracts with coal mines, because of a desire to invest in nuclear power plants. As a result, coal mine owners stopped opening new mines and signed contracts to export much of the coal they did mine to countries outside the United States.

Besides the sulfur and nitrogen oxides, fossil-fueled plants also produce soot and fly ash, and large quantities of carbon dioxide and water vapor. If fossil fuel plants make up half the power production up to the year 2000, and if power production continues to rise at the present rate, it is estimated (13) that by the end of the present century, the CO_2 from coal-burning plants will amount to 8.75 billion tons per year, some 0.5 percent of the total carbon dioxide in the atmosphere. Water vapor, produced by both fossil- and nuclear-fueled plants, would also be added in large amounts, and might very well affect world climate within our lifetimes (13).

Besides the more obvious air pollution hazards from combustion of fossil fuels, there is the problem of the environmental cost of mining the coal in the first place. Acid mine drainage is a serious problem, and some of the worst environmental insults in man's history have come from strip mining. Not only is the land eroded and ruined, but water running over these shallow exposed coal mines carries away sulfuric acid, which ends up in our streams and rivers. Deep coal mining, on the other hand, presents well-known hazards to the health of miners. Of course, the mining and transporting of oil also has its environmental hazards, so it would seem that environmental costs, although up to now less obvious to many of us than the technology behind the generating of electricity, are unavoidable. Let us examine the picture using other than fossil fuel.

Unique Features of Nuclear Power Plants

Nuclear plants are still relatively new and still experimental forms of power production. By the end of 1970, some 19 civilian nuclear power plants had been completed in the United States, since the first commercial nuclear plant became operational in 1957 at Shippingport, Pennsylvania. Although these account for less than 2 percent of the power used in 1970, some 54 more nuclear plants are now being built, and it is estimated by the Atomic Energy Commission that by 1980 approximately 30 percent of our electrical generating capacity will be nuclear. By the end of 1968, however, five of the 19 nuclear plants had been shut down as impractical or unsafe, and a sixth, the Fermi reactor outside of Detroit, never operated properly; it suffered a nearly disastrous accident on October 5, 1966, and was never able to resume normal operation. It is now being dismantled. The Fermi reactor differed from most of the other types in that it was designed as a so-called "breeder" reactor, a type to be discussed in the following section.

Although nuclear-fueled power plants have the distinct advantage of producing none of the noxious and sometimes poisonous air pollutants that fossil plants do, they produce their own brand of pollution. Not only do they release tiny amounts of radiation during normal operation, and the waste heat hot water discharges we have already mentioned, but they produce radioactive wastes that must ultimately be disposed of. Finally, in case of accident, they could unleash clouds of deadly radiation. The day-to-day low-level radioactivity from nuclear reactor operation comes principally from two radioisotopes: tritium (hydrogen-3) and krypton-85. All other fission products can, barring accident, be relatively easily contained in the spent fuel, and theoretically buried harmlessly. Tritium and krypton-85, however, produced in approximately 1 and 30 of every 10,000 fissions, with half-lives of 12.4 and 10.4 years, respectively, present special problems. Tritium is chemically identical to ordinary hydrogen, so it can escape in the chemical form of water and circulate within the world's water supplies. Krypton-85, an inert gas, can escape into the air, so both these isotopes are difficult to contain once a spent fuel element is sent to a chemical reprocessing plant.

Because of the potential danger of any amount of radioactivity, an advisory committee of the National Academy of Sciences specified that all radioactive wastes be isolated from the biological environment, that waste disposal practices be designed for the much higher levels of waste production likely in the future, and that safety not be compromised for the sake of economy (12). While highly radioactive wastes can best be isolated by converting them to solid ceramic-type chemical compounds and then burying them in natural salt beds which are highly impervious to groundwater flow, this is not always done. Lower level wastes are often buried in trenches 10 to 15 feet below ground, where soil moisture could return radioactivity to the surface by evaporation, or where it could descend to the underground water table (12). Thus, it is quite apparent that the advisory committee recommendations have often not been followed with regard to radioactive waste disposal. The continuation of such practices in the future, when these wastes reach 10 to 100 times their present volume, could cause serious hazard.

The mechanics of controlled, reactor-type nuclear fission was illustrated in Fig. 4–14B. Initially, all nuclear fission depends solely on the U-235 isotope, which comprises only 0.7 percent of naturally occurring uranium. The bulk of this natural uranium—almost 99.3 percent—is U-238, with a third isotope (U-234) present only in negligible trace amounts (0.006 percent). The significance of U-235 is due to the fact that it is the only naturally occurring isotope that fissions spontaneously by capture of slow neutrons. We have already indicated that the known supplies of natural uranium, when projected against future worldwide power needs, would be exhausted in less than 20 to 25 years if used as the principal industrial fuel (12). This sobering fact has led to the development of several types of reactors, which can be conveniently divided into three principal categories: burners, converters, and breeders.

Fast-Breeder Reactor Nuclear Power

It has been well known, since the development of atomic energy in the early 1940s, that the U-234 isotope fissions when absorbing neutrons (Fig. 4–13A), while the more abundant U-238 simply absorbs the neutrons and is transmuted, ultimately, to plutonium (Fig. 4–12B). The isotope Pu-239, like U-235, undergoes spontaneous fission upon absorbing neutrons. Taken together with the fact that each such fission releases 2 to 3 potentially usable neutrons, some interesting possibilities suggest themselves. One is simply that U-235 might be separated from U-238 and allowed to undergo fission, producing useless fission products, and simultaneously producing heat that could be used, ultimately, to generate electricity. Such reactors are known as "burners," since they simply burn up their uranium-235 fuel. Most currently used electric power reactors are of this type, or the type known as "converters." A converter contains *both* U-235 and U-238 or U-235 and thorium-232 (Th-232), which comprises the bulk of naturally occurring thorium deposits. Both Th-232 and U-238 can then absorb some of the neutrons released during U-235 fission, since only 1 of the several neutrons released per fission is needed to keep the fission chain reaction going in the U-235. These extra neutrons thus absorbed can convert non-fissionable U-238 to fissionable Pu-239 (Fig. 4–12B), and, in a similar manner, non-fissionable Th-232 into fissionable U-233.

This means it is theoretically possible for a nuclear reactor to make new fissionable fuel while it is consuming its original fissionable U-235 fuel. Reactors which produce as much or less fuel than they consume are known as "converters," while those that produce *more* fuel than they consume are called "breeder" reactors. It is the breeder that is thought by nuclear chemists to hold the greatest hope of eventually solving our fuel-shortage problems. Because they can at least theoretically produce *more* fuel than they consume, they could utilize the enormous quantities of low-grade uranium and thorium ores dispersed in the Earth's rocks as a source of low-cost energy for thousands of years (15).

Besides the fuel (U-235 initially), most reactors employ a control system of boron rods that can be inserted into the reactor to absorb neutrons released by fission, thus preventing them from fissioning additional U-235 atoms. By inserting such control rods among the fuel elements, a fission chain reaction can be slowed or stopped. If the control rods are withdrawn, the chain reaction can be started or increased. Besides control, reactors usually have a coolant and moderator. The coolant is a fluid which circulates around the fuel elements and carries away the intense heat produced during fission. Most United States reactors are cooled by water, but some use a gas as a coolant. The moderator is often employed to slow down neutrons released during fission, because slow-moving neutrons are more likely to produce fission than fast-moving ones. Usually, coolant water can also act as moderator.

The most promising type of breeder, however, is one that uses liquid sodium metal as a coolant. The reactor is contained in a large tank of liquid sodium, and separated from the primary heat exchangers and the associated pumps by loops of piping through which sodium coolant flows. While sodium is not as effective as water in slowing down fission neutrons, it has been found that this is an advantage when we wish to breed more fuel quickly, because while fast neutrons do not easily cause fission, they are effectively absorbed by U-238 atoms also present in the fuel, converting these to fissionable Pu-239. It turns out that this conversion utilizes more fission neutrons, wasting less than when the neutrons are slowed by a moderator, and therefore breeding more Pu-239 fuel more quickly than any other type of breeder now known. This type is known as a liquid-metal-cooled fast-breeder reactor, and may be able to produce twice as much fuel as it consumes every 7 to 10 years. In other words, starting with a given amount of U-235 fuel, mixed in with lots of U-238, such a reactor would "breed" enough Pu-239 (Fig. 4–12B) in 7 to 10 years to fuel two reactors similar to the original breeder (15).

It must be emphasized, however, that breeder reactors are highly experimental at present, and the most optimistic predictions are that they will not be used to generate commercial electric power before 1984 (15). The prototype liquid-metal Fermi reactor, as previously mentioned, suffered a very nearly serious accident in 1966, when the liquid sodium coolant flow was momentarily blocked, and this resulted in the melting of some of its uranium fuel. The possibility of such accidents causing large-scale release of radioactivity is a real one, even if the probability is low. The sodium coolant is highly reactive and

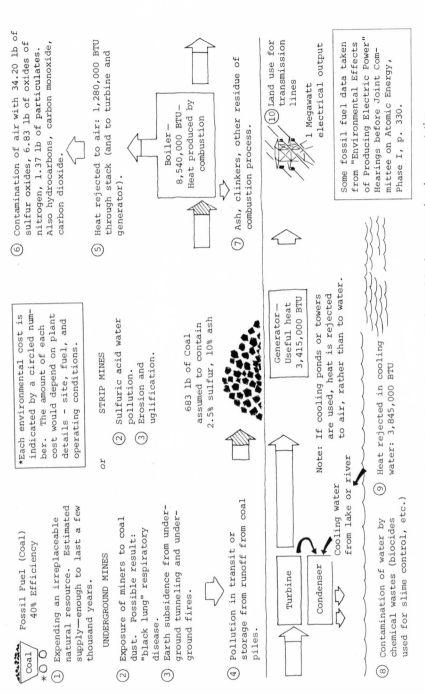

Fig. 5–10. Environmental cost of producing 1 megawatt of electricity in a steam-electric power station using fossil fuel (coal). From Abrahamson (13).

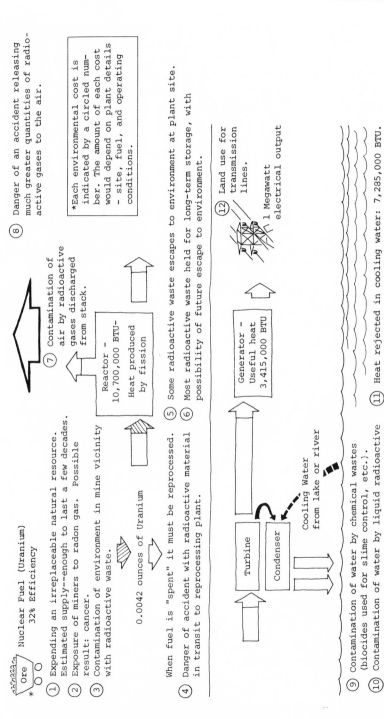

Fig. 5–11. Environmental cost of producing 1 megawatt of electricity in a steam-electric power station using nuclear fuel (uranium). From Abrahamson (13).

Nuclear Fuel (Uranium)
32% Efficiency

① Expending an irreplaceable natural resource.

② Estimated supply--enough to last a few decades. Exposure of miners to radon gas. Possible result: cancer.

③ Contamination of environment in mine vicinity with radioactive waste.

0.0042 ounces of Uranium

④ When fuel is "spent" it must be reprocessed. Danger of accident with radioactive material in transit to reprocessing plant.

⑤ Some radioactive waste escapes to environment at plant site.

⑥ Most radioactive waste held for long-term storage, with possibility of future escape to environment.

⑦ Contamination of air by radioactive gases discharged from stack.

⑧ Danger of an accident releasing much greater quantities of radio-active gases to the air.

*Each environmental cost is indicated by a circled number. The amount of each cost would depend on plant details - site, fuel, and operating conditions.

Reactor -
10,700,000 BTU-

Heat produced by fission

Turbine

Condenser

Cooling Water from lake or river

Generator -
Useful heat
3,415,000 BTU

1 Megawatt electrical output

⑫ Land use for transmission lines.

⑨ Contamination of water by chemical wastes (biocides used for slime control, etc.).

⑩ Contamination of water by liquid radioactive waste.

⑪ Heat rejected in cooling water: 7,285,000 BTU.

dangerous if exposed to moist air, adding to the risk of such reactors. It seems apparent, even if all radioactivity can be contained in the breeder, and no accidents occur, that the reprocessing, transporting, and burying of the highly radioactive fission products will become an ever-increasing problem. The more electricity that is generated by nuclear power, the more such fission products will be created, and have to be disposed of. Figures 5–10 and 5–11 compare the environmental effects of fossil-fuel and nuclear-fuel generation of electricity.

Energy from Nuclear Fusion Reactions

The hydrogen bomb reaction was discussed in Chap. 4 and illustrated in Fig. 4–13B. Ever since Hans Bethe worked out the sequence of nuclear reactions by which such enormous amounts of energy are produced in the Sun and stars, and his results were published in 1939, the question of whether such controlled fusion-type reactions could be achieved in the laboratory has been a continuing challenge. The most hopeful approaches to this challenge to date are those involving the fusion of 2 deuterium (hydrogen-2) atoms, or of 1 deuterium and 1 tritium (hydrogen-3) atom; the second approach appears more likely to be successful.

We have already seen how a compound of lithium-6 deuteride, $^6_3Li^2_1H$, can react with neutrons from an atomic bomb to produce tritium, and how the tritium and deuterium can fuse at the temperatures produced in an atomic explosion. First, large quantities of lithium-6 and deuterium must be made available. The deuterium, occurring naturally, is found in all water with a relative abundance of 1 deuterium atom for each 6,500 hydrogen atoms. This may seem to be a small number, but considering the large amount of seawater available, the deuterium could be separated in sufficient quantities. Lithium, on the other hand, is found in readily extractable concentrations only in restricted localities on land. In seawater, it is present in only 1 part in 10 million (12), and its abundance in the Earth's crust is not much greater. In addition, the necessary isotope, lithium-6, constitutes only 7 percent of natural lithium. It has been reliably estimated that the amount of lithium-6 potentially available in North America and Africa is only one-hundred-millionth of the amount of deuterium extractable from seawater (12). This means the lithium-6 supply will determine the amount of energy potentially available from the type of fusion reaction we have been discussing, and it has been calculated that this amount of energy is roughly equal to that obtainable from combustion of the world's initial supply of fossil fuels (12).

Thus, this most promising lithium-6 deuteride reaction may not be as practical to pursue as the seemingly more difficult to achieve deuterium-deuterium fusion. The problem of how one achieves the high pressures and temperatures necessary for fusion to occur, regardless of the fusion reaction used, is another area that has not been resolved. The approach thus far has been to construct so-called magnetic bottles, inside of which deuterium and tritium are raised to very high temperatures by massive jolts of electric power. These substances must be contained by powerful magnetic fields, and the goal of containing them until their temperatures reach the millions of degrees needed for fusion to occur has thus far not been realized. Even if these seemingly insoluble problems were somehow solved, and a working thermonuclear (fusion) reactor some day built, there would still be the danger that some of the highly radioactive tritium would escape and enter our water supply. According to fusion reactor design estimates, some 15 percent of the tritium produced in the lithium-6 neutron reaction is expected to be lost along the way (16), which may amount to 20,000 times the amount of tritium released by generation of an equivalent amount of electricity by a fission reactor. Under the circumstances, it seems safe to conclude that the outlook for fusion power is far from rosy at the present time.

Other Sources of Energy: Solar and Water Power

The most obvious energy source is the Sun, for it has been estimated that the amount of the Sun's heat intercepted by the Earth is about 100,000 times larger than the world's present installed electric power capacity (12). For a solar electric power plant of 1,000-megawatt capacity (that of large modern power stations today), calculations indicate that such a plant would require a square area of about 4 miles on a side! This assumes the most reliable current estimates that (*1*) there would be a 10 percent conversion from solar to electric power, and (*2*) an average solar power input at the Earth's surface of 2.4×10^{-2} watts/cm^2. The calculation is then a straightforward one: to produce 1,000 megawatts ($1,000 \times 10^6$ watts, or 10^9 watts), 10 times this amount must be collected. Hence:

$$\frac{10^{10} \text{ watts}}{2.4 \times 10^{-2} \text{ watt/cm}^2} = 42 \times 10^{10} \text{ cm}^2 = 42 \text{ km}^2$$

Since 1 km = 0.62 miles, this amounts to 42 km^2 $\times$ [(0.62 mi)^2/1 km^2] $\doteq$ 16 mi^2, or 4 miles on a side (12). At present, therefore, it appears impractical to utilize solar energy, since doing so would tie up so much of the land.

Water power represents the largest concentration of solar power produced by any natural process, for clearly it is the Sun's heat that evaporates water from the oceans, allowing it to fall as rain and thus form natural reservoirs above sea level (see Fig. 2–2). Five hydroelectric plants already exist in the United States, each with power capacities exceeding 1,000 megawatts (12), in which the motion of the water flowing to lower levels is used as the source of mechanical motion. This represents less than 20 percent of the total potential in the United States, which raises the question of why more should not be done in this area. The U.S. Federal Power Commission estimated in 1966 that the total potential water-power capacity of the world amounted to some four times the amount of its total installed electric-power capacity (12).

There are two environmental considerations that offset the obvious advantage of producing electric power free from air or radioactive pollution. The first is the aesthetic consideration of sacrificing the natural beauty of the areas to be developed; this would occur even when natural waterfalls are used for power generation. The majority of our hydroelectric projects involve the damming of large amounts of water which is then used to generate electric power. This leads to the second drawback: All reservoirs formed by damming rivers or streams are at best temporary, for these streams are continuously depositing their loads of sediments. Thus, in a century or two, the reservoirs become completely filled with such sediments, and hence useless. Associated ecological effects are often difficult to predict, although they may turn out to be quite serious. Egypt's Aswan Dam, for example, designed around goals of power production, irrigation, and flood control, has produced undesirable effects as well. Its restriction of the flow of silt down the Nile has prevented the silt from enriching the lands along the Nile's banks to help offset their natural erosion. Thus, downstream erosion may wash away more productive farmland than will be opened up by new irrigation systems near Lake Nasser. Also, this silt, formerly reaching the Mediterranean and fertilizing organisms which support fish life, has by its absence been largely responsible for the decline of the Egyptian sardine catch from 18,000 tons in 1965 to 500 tons in 1968 (13).

Yet another effect of the Aswan Dam, also observed in an older British irrigation and power project in Pakistan, may prove to be the most serious of all. It is that the underground water level of the entire region may be raised by the damming of large amounts of water, and that this, in turn, may dissolve salts and other minerals in the subsurface, bringing them close enough to the productive soil to make farming of it difficult or impossible (13). Thus it is apparent that such projects may well upset natural balances to the point where power gains are more than offset by the negative effects of such large, man-made dams.

Future Power Prospects

Our examination of energy requirements reminds us that the production of ever-increasing amounts of electric power is not without its environmental cost. If we are to keep these costs within acceptable limits, we must challenge the notion that electric power production can continue to increase at its present rate: that of doubling every 10 years. When we examine the anatomy of this future demand, we find that relatively little demand comes from private consumers, for home electric heating and air conditioning. Most electric power goes, and will continue to go, to industry and to commercial use, so it is these sources that must, along with the consumer, look for ways to limit demands for electric power. Primary metals industries, especially aluminum, along with automobile manufacturing, are the largest consumers of electricity (13). Yet there is no valid reason why we cannot return to using glass instead of aluminum beer containers, or build automobile bodies to last instead of needing to be junked and replaced every several years. In addition, we must explore more efficient and, in some cases, new ways of generating our electric power needs.

One obvious but neglected source is power from garbage. Municipal solid wastes, an ever-increasing problem to dispose of anyway, constitute a better fuel than is generally recognized. They have an average heating value, when burned, of over 1,200 kilocalories per pound, which is about one-third that of a good grade of coal (13). Paris, France, began solving its garbage problems by burning refuse for power over 50 years ago. Today, over 1.5 million tons of garbage is collected each year from Paris and its suburbs, and three-fourths of it is converted into electric power in four big plants and sold to the city to heat hundreds of Paris buildings (13). In the United States however, such plants are recent and rare, and their development needs to be studied and encouraged.

Magnetohydrodynamics (MHD), already tested in prototype form, uses coal in a way that is twice as efficient as conventional systems, by converting the coal to a gas, seeding it with millions of tiny particles to make it more conductive, and forcing it through a stationary magnetic field. Thus, electric current is produced directly (it is direct current), pollution is less because combustion is more complete, and exhausts are recycled to recover the seeding material. The present limitation has been the inability to find structural materials that can stand the high temperatures ($4,000°-5,000°F$) that accompany the process, creating doubt that MHD could be counted on to supply much electrical energy before the year 2000 (13).

A more promising future power source is the Earth's interior heat. Indeed, Pacific Gas and Electric currently operates such a geothermal generator some 90 miles north of San Francisco, where it is expected to be generating 600 megawatts of electricity by 1975 (17). The source of geothermal energy is the molten rock, or magma, in the Earth's interior. When underground water comes into contact with the magma, hot water and steam are produced. If this happens relatively close to the Earth's surface, as it does in California's Imperial Valley, it becomes possible to tap and pipe the steam and hot water to the surface, where it is used to heat a gas such as isobutane. The heated isobutane can then be used to turn turbines, while the hot water is pumped back into the underground reservoir. Such a heat transfer system, perhaps more complex than using the steam directly, eliminates two possible sources of environmental damage: left-over brine, brought up with the hot water, which might be toxic and hard to dispose of, and possible sinking of the ground caused by removing the subsurface water. Besides being much less polluting than more conventional methods of producing electricity, there is also the possibility, not yet fully confirmed, that such geothermal energy might be a self-renewing resource (17).

The Pesticide Problem

We have already traced the development, wide use, and harmful effects of chlorinated hydrocarbon pesticides such as DDT, as well as the more recently developed organophosphates. And we have seen how difficult it is for people to give up using an insecticide, even after they know of its dangers. Even today, pesticides are widely sold and used in spite of their potential or well-known dangers. Probably the best-known example of this is the currently popular "No-Pest Strip," made by the Shell Oil Company, a subsidiary of the international Royal Dutch Shell Oil Company (18). This is the familiar 10-inch plastic strip, used by suspending it within its oblong paper cage, and releasing by slow evaporation the newly developed pesticide DDVP. According to its advertising, the strip releases this pesticide continuously over a 3-month period (18).

DDVP, short for O,O-dimethyl-2,2-dichlorovinyl phosphate (Fig. 5–12), was tested in 1955 by the U.S. Public Health Service during a search for chemicals which could be used to kill insects aboard airplanes in international flights, and to control flying insects in areas of the world where such insects posed serious public health problems and were becoming resistant to DDT. DDVP was never intended for wide use in United States homes, where no such public health problems exist. According to *Environment Magazine* (18), there

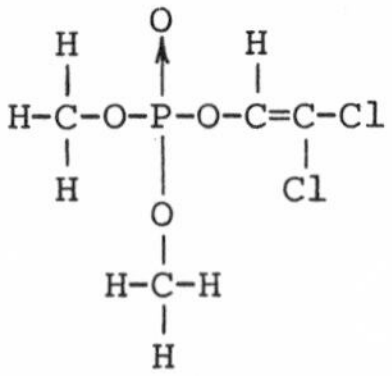

```
        O
   H    ↑    H
   |    |    |
H-C-O-P-O-C=C-Cl
   |    |    |
   H    O    Cl
        |
      H-C-H
        |
        H
```

Fig. 5–12. The organophosphate insecticide DDVP, short for *O, O*-dimethyl-2,2-dichlorovinyl phosphate, the active ingredient of Shell Oil Company's "No-Pest Strip."

has never been a study of the effects of inhaling the pesticide steadily over a period of years, although this is just what is happening in numerous American homes and public buildings. Indeed, No-Pest Strips are registered by the Department of Agriculture for use in homes and institutions, in spite of the lack of studies of the long-term effects. DDVP is a member of the organophosphate family of chemicals, related to the nerve gases developed during World War II.

Both the nerve gases and organophosphate insecticides such as DDVP apparently attack a vulnerable portion of the nervous system. Normally, the chemical acetylcholine assists in carrying impulses from one nerve fiber to the next, and from nerves to muscles. The enzyme acetylcholinesterase breaks down this messenger chemical after it has served its purpose. Anything which interferes with the delicate balance between acetylcholine and acetylcholinesterase severely disrupts the nervous system. Organophoshate insecticides and nerve gases do this, it seems, by preventing the acetylcholinesterase from breaking down the acetylcholine. This results, in man and other mammals, in a buildup of acetylcholine, accompanied by a barrage of extraneous nerve impulses which disrupt bodily functions and eventually lead to death by suffocation because of the paralysis of the diaphragm (18).

Thus, people who inhale DDVP while ridding themselves of insect nuisances are simultaneously exposing themselves to a material which, in large doses, is known to be toxic to man. The obvious question of whether it is safe to keep small quantities of DDVP in the air—enough to kill flies but not enough to harm humans—has apparently never been answered. It should have been answered before permitting DDVP-vaporizing devices, such as the No-Pest Strip, to be so widely accepted into general use.

A similar but more disturbing example of such carelessness has been the continuing sale and use of Lindane-containing insect vaporizers. Lindane (Fig. 4–6) belongs to the DDT-type of chlorinated hydrocarbon insecticides, and it affects the nervous system of insects and higher animals in an as yet undetermined way. In human beings, exposure to Lindane vapors has been linked to conditions ranging from mild skin reactions to a very serious blood disorder known as aplastic anemia (19). Although its potential hazards have been known since 1951, the effects of continued inhalation of Lindane vapors for long periods has never been tested (19). In spite of continued warnings by scientists, it was not until 1969 that the Department of Agriculture finally canceled its registration of Lindane for use in the home.

The alarming aspect of the Lindane vaporizer episode is that such devices are still being purchased in hardware stores in the United States, and they are sold without proper warning of their potential hazard. As early as 1954, for example, the American Medical Association's Committee on Pesticides warned that, in spite of laboratory and clinical evidence of Lindane vapor toxicity, promoters continued to represent their appliances as absolutely safe (19). Even though the Department of Agriculture has canceled its Lindane registration, Lindane vaporizers are still being sold with enough Lindane pellets to be clearly toxic if accidentally swallowed. The only warning, the federally required "For Commercial or Industrial Use Only, Not To Be Used in Homes or Sleeping Quarters," was found by *Environment Magazine* investigators at the bottom of a page of general directions printed on one side of the packet of Lindane crystals. No such warnings appeared on the vaporizer package itself (19).

Thus the consumer is left virtually unprotected, with eight manufacturers still marketing Lindane vapor dispensers over a year after federal authorization was canceled, simply because they have appealed the cancellation (19). Clearly, the federal authorities must be pressured into stricter enforcement, and the registration laws themselves must be amended to prohibit the flouting of government attempts at protection by simply appealing decisions to cancel registrations. In the meantime, the consumer must beware, must take pains to discover just what he is being sold, and not depend on product claims and advertising to do this job for him.

Analysis and Control of Pesticide Residues

The widespread use of pesticides has already been examined, and some alternatives to such continued use must now be explored. What can be done, however, with persistent pesticides such as DDT that are already widely distributed in the environment, and constitute a threat to it? First, we must be able to monitor the extent and concentrations of such residues accurately. Scientists have become keenly aware of the problems inherent in such monitoring, and are attempting to evolve better methods (20).

Gas chromatography, a relatively new analytical technique (proved feasible in 1952), provides the most reliable method presently known for detecting the presence of pesticide residues. While gas chromatographs capable of such analyses are expensive and not likely to be found in high school laboratories, the principle behind their operation is easily understood. Each chemical compound has a different rate of adsorption onto and desorption from some filter-type material, which means that each compound will take a different length of time to travel through a column packed with such material. In practice, a gas chromatograph has three main sections: a flash heater to vaporize the material to be analyzed, a separation column containing the filter material, and a detector to scan the gas mixture as it leaves the column. A carrier gas, usually nitrogen or helium, is used to "push" the vaporized material through the column.

The problem of how to induce the breakdown of persistent chlorinated-hydrocarbon-type pesticides into simpler substances less harmful to the environment is now under intensive study. Despite the enormous concern with the persistence of DDT, surprisingly little is known about the decomposition of DDT-type pesticides by microorganisms present in the natural environment. The first priority, therefore, would seem to be the need for more information about the transport and fate of pesticides, out of which some possible solutions may suggest themselves.

Food Crops and Insect Control

We have already mentioned the pattern of increasing insect pest populations actually being created by our increasing food crop production, and it has led ecologists such as Lamont C. Cole to warn that clean cultivation, routine pesticide application, and other agricultural practices that reduce the diversity of species in the community may be working in exactly the wrong direction. This is because a healthy, diverse biotic community is not easily invaded, and has considerable ability to adjust to invaders (21). Dr. Cole points out that medical researchers have concluded that it is more desirable to replace harmful viruses and bacteria with innocuous types, rather than leave niches open by trying to keep our bodies free of these forms. Since a broad-spectrum anitbiotic or pesticide (DDT, Lindane, DDVP) is likely to empty several niches because it is not specific for one organism, but kills many, both antibiotic and pesticide may produce the same effect. Ever since antibiotics came into use, doctors have been plagued by secondary infections resulting from the destruction of a harmless intestinal bacterium, thus leaving ecological niches available for drug-resistant and dangerous germs (21).

The same principle applies to biotic communities, Dr. Cole believes, and he cites examples such as secondary infections due to orchard mites replacing the codling moth as the pest of apples, and spider mites replacing the spruce budworm as the defoliator of fir trees, both following insecticide application. The insects that have succeeded in adapting to DDT or other man-made insecticides, while their weaker relatives and natural enemies are often killed, reach populations far greater than those they could achieve in the absence of the insecticide. Doctors now, by and large, treat a patient only when treatment is needed and, by preference, will use the most highly specific drug available for the particular infectious organism (21). It seems perfectly obvious that truly effective long-term insect control can be achieved only by analogous specific methods. It was most discouraging to learn recently, so many years after Rachel Carson documented the damage caused by the Department of Agriculture's previous attempts to eradicate fire ants in the South, that the Department was trying to repeat the fiasco. In the late 1950s, Heptachlor was used, and discontinued only after traces of it began to appear in the milk of cattle. Plans for a 12-year, $200 million program against fire ants call for using the chemical Mirex, which, like Heptachlor, is a cyclodiene, and is a cancer-causing agent as well as a broad-based poison; these plans have caused much concern among ecologists (22), and demonstrate the continuing need for public awareness and pressure against such short-sighted programs.

Alternatives to the Use of Conventional Pesticides

While man must control insect pests because they compete with him for the means of survival, he must begin to be more circumspect about the means of control. It is now recognized that the method of combating a pest or disease should be specific to the target organism, interfering only with its welfare, and that the method should not introduce new contaminants into the environment (23). The development and growing of crops that are immune or resistant to pests and diseases is one obvious way to do this, although relatively little progress has been made in developing plant resistance to specific insects (23). Another effective method would be increased use of biological controls, ways to use insect parasites, predators, and diseases to prey upon damaging pests and thus control their numbers. Recent tests have shown, for example, that the release of 200,000 aphid lions per acre for a sustained period was as effective against the bollworm as the available insecticides (23); another study indicated that 100 million parasitic wasps released on 18,000 acres of alfalfa effectively controlled pea aphids by suppressing their populations and delaying their migrations to pea plantings (23).

The potential value of such biological controls is promising, for parasites are usually specific in action, and thus far have had no damaging effects on the environment (23). Much more work would be desirable to overcome the complex obstacles of finding and mass-producing suitable biological agents.

The manipulation of insects for their own destruction, by inducing sexual sterility or introducing other harmful traits, is a relatively new approach to insect control, and one which holds considerable promise. Two distinct methods of using sterility as a control are being studied. One method is based on the rearing of massive numbers of a pest species, sterilizing them with gamma-radiation, and releasing the insects to compete for mates in the natural population. The resulting eggs do not hatch, and the insect population dwindles. The second method involves the application of chemosterilants to native populations at a central source. The treated insects then disperse and serve to reduce the reproduction of target pests in the environment (23). Both methods have been used in some cases with considerable success.

Insect Attractants

One of the newest trends is research to identify and develop attractants and hormones for insect control. Scientists are investigating insect responses to various chemical substances in the plants the pests feed upon, to chemical sex attractants, to light, and to sound. Naturally occurring attractants are highly specific and active in infinitesimal amounts, and intensive effort is being devoted to the isolation, identification, and synthesis of several sex-related chemicals so as to obtain sufficient amounts for practical use in the control of important pests (23). The first sex attractant (also called a pheromone) was isolated from the female gypsy moth in 1960, and since then some 200 have been discovered.

The promise of an approach that involves the use of odorous chemicals to lure insects into lethal traps lies in the fact that many insects find their food, their partners for mating, and favorable sites in which to deposit their eggs by means of automatic responses to various scent cues. Male moths, for example, can smell potential sexual partners at a considerable distance. The specificity of this approach is due to the tendency of each insect species to have its own distinctive odor, for this facilitates the meeting of partners capable of mating with each other. Thus, instead of spraying the whole countryside with insecticide, one could attract the unwanted insects to traps where they would contact an appropriate insecticide, a sterilizing chemical, or some other exterminating device. In addition, females might be induced to lay their eggs not on nourishing plants but in places where the emerging larvae would starve for lack of food. Such specific techniques promise to be much less costly and much less environmentally damaging than the widespread use of insecticides (20).

The investigation of insect attractants goes back to discoveries made in the nineteenth century by Jean Henri Fabre, the French naturalist, who found that a female moth attracted male moths only when placed in an open cage, rather than in an airtight, transparent container in which the female could be seen but not smelled. He also found that the open container recently occupied by the female was still attractive to males, even though the female was no longer there (20). It has since been discovered that male gypsy moths are attracted to a female up to half a mile away; in some cases, reception of just a few hundred molecules or less of the female scent by the sensitive cells of the antennae is enough to stimulate the male (20).

The principle difficulties in insect attractant work are the problem of identifying the specific odorous substances to which various pests respond, and, in the cases of the natural sex lures, the task of obtaining large enough amounts of these substances to carry out large-scale programs. To concentrate just 12.2 mg of pure female cockroach sex attractant, for example, took experimenters 9 months and 10,000 females. The first use of sex attractants as a weapon against insects was against the gypsy moth, first brought to the United States in 1869 in a futile attempt to start a silk-growing industry. Some of the species escaped, multiplied, and became a major New England tree pest. Since the female cannot fly, she depends on the male's attraction to her scent. After a long search, the gypsy moth sex attractant was finally identified (Fig. 5–13) by Department of Agriculture scientists in 1960, but it took the abdomens of half a million female moths to yield 20 mg (less than one-thousandth of an ounce!) of the pure attractant (20). A trap baited with one ten-millionth of a microgram of the attractant would lure males to it!

The specificity of attractants, each luring only its own species, is remarkable. When the gypsy moth attractant discoverers were able to synthesize the chemical in their laboratory, they couldn't resist attempting to improve on nature by synthesizing related compounds they hoped would be more effective.

Fig. 5–13. Insect sex attractants. Note that the two forms of the Siglure isomers differ only in the arrangement of groups around two adjacent carbons. The *cis* form has the methyl and ester groups on the same side of the ring, the *trans* on opposite sides. The medfly much prefers the *trans* form, both of Siglure and Trimedlure, a more powerful attractant synthesized by adding hydrogen chloride to the double bond in *trans*-Siglure. From Jacobson and Beroza (20, 24) and Edward O. Wilson, "Pheromones," *Scientific American*, Vol. 208, No. 5, May, 1963, p. 108.

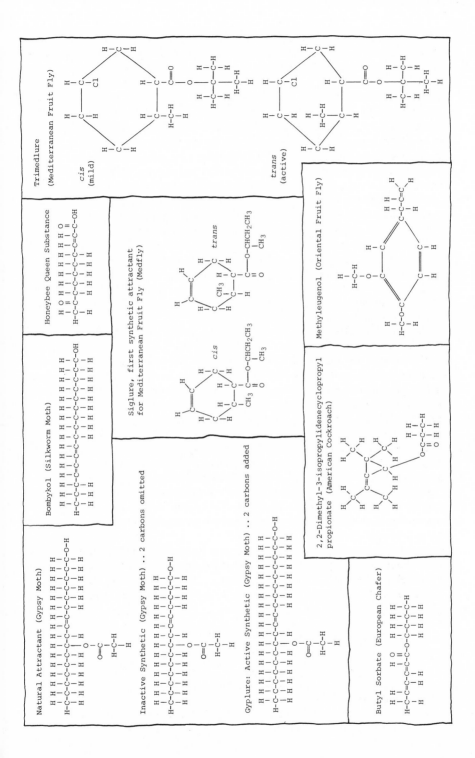

They felt that reducing the size of the molecule, from 16 to 14 carbon atoms in chain length (Fig. 5–13), might, by making a more volatile compound, make it more attractive. This shortened molecule failed to attract the males, but a product in which the carbon chain was lengthened by 2 carbon atoms (Fig. 5–13) was found to be just as attractive to males as the natural attractant, and much easier to synthesize in large quantities. Called Gyplure, it has since been adopted by the Department of Agriculture as the agent for trapping gypsy moths, and is made from an ingredient of castor oil. Its use has been credited with preventing the spread of the gypsy moth outside the New England area, while the use of the subsequently discovered Mediterranean fruit fly attractant helped to eradicate this pest from infested areas of Florida (20) just a year after it was first discovered there in 1956.

While sex attractants hold great promise for future insect control, their use through the early 1970s has been primarily as an early warning system that enables man to discover and hopefully eradicate an infestation of a foreign species (such as the Mediterranean fruit fly) before it can become widespread (24). A trap baited with a particular attractant that catches the responding insect signals the presence of its species, and hence indicates when and where that species must be fought—with the best available methods (24).

Insect Hormones

A relatively new approach to insect control has been suggested by basic studies of insect physiology, particularly the juvenile hormone that all insects secrete at certain stages in their lives. It is one of the three internal secretions used by insects to regulate growth and metamorphosis from larva to pupa to adult; at certain stages each hormone must be secreted, and at other times it must be absent or the insect will develop abnormally. The immature larva, for example, needs juvenile hormone to progress through the usual larval stages, but the flow of hormone must stop if the mature larva is to metamorphose into a sexually mature adult (25). Still later, after the adult is fully formed, juvenile hormone must again be secreted, while the hormone must be absent from insect eggs for the eggs to develop normally (25).

The times at which the hormone must be absent may be when the insects are most vulnerable. If the eggs or insects come into contact with the hormone at these times, the hormone readily enters them and provokes a lethal derangement of further development. The result is that the eggs fail to hatch or the immature insects die without reproducing (25). Since juvenile hormone is an insect invention that has no known effect on other forms of life, it should be able to zero in on pest insects to the exclusion of other plants and animals. Indeed, it does not even kill the insects, but derails their normal development and hence causes the insects to kill themselves. Furthermore, insects will not be able to evolve a resistance or insensitivity to their own hormone without automatically committing suicide (25).

The potentialities of juvenile hormone as an insecticide were first recognized in 1955 in experiments performed with the first active preparation of the hormone, a golden oil extracted with ether from male cecropia moths. Indeed, the male cecropia and its close relative, the male cynthia moth, are essentially the only insects from which man has yet learned to extract the hormone (25). Since the insects themselves were the source of microscopic amounts of the hormone, it would have to be synthesized in quantity in the laboratory to be available for use as an insecticide. This could not be done, however, until the hormone had been isolated from the golden oil and identified.

This was finally done, in 1967, by Herbert Röller of the University of Wisconsin, and the hormone was found to have the empirical formula $C_{18}H_{36}O_2$, corresponding to a molecular weight of 284. It is the methyl ester of the epoxide of a previously unknown fatty acid derivative (Fig. 5–14A), and is less simple than it looks. It has two double bonds and an oxirane ring (the small triangle at the lower left of the molecular diagram), and can exist in 16 different molecular configurations, only one of which is the authentic hormone (25). Tests conducted at the University of Wisconsin indicate that the hormone is extremely active, 1 gm of it resulting in the death of perhaps a billion insects! This insect-produced hormone, however, with two ethyl groups $(CH_2 \cdot CH_3)$ attached to carbon atoms 7 and 11 (Fig. 5–14A), is virtually impossible to synthesize from any known chemical compound. Department of Agriculture chemists were able to synthesize an analogue of this juvenile hormone (Fig. 5–14B) which differed from the authentic cecropia hormone by only 2 carbon atoms. It is only 0.02 percent as active as the pure cecropia hormone! Some years later, by bubbling hydrogen chloride gas through an alcoholic solution of a chemical known as farnesenic acid, chemists in the United States prepared a mixture fully as effective as the natural hormone. A Czechoslovakian chemist was able to isolate and identify one of the components of this mixture (Fig. 5–14C), and estimated that from 10 to 100 gm of it would clear all the insects from 2½ acres of infested land (25).

All the materials thus far mentioned are selective in the sense of killing only insects, and do not therefore discriminate between the 0.1 percent of insects that qualify as pests and the 99.9 percent that are helpful or innocuous. Therefore, any reckless use of this type of hormonal material might produce an ecological disaster. The obvious need is for chemicals that are tailor-made to attack only the target pest insect, leaving others unharmed. A strange accidental discovery has recently indicated that this is not an impossible goal. It seems that some specimens of the European bug *Pyrrhocoris apterus*—a species that had been reared in a Czechoslovakian laboratory successfully for over 10 years, mysteriously died without reaching sexual maturity when Harvard University scientists attempted to breed them. Analysis of this incident suggested some juvenile hormone effect. Eventually the problem was traced to the paper toweling that had been placed in the insect-rearing jars, and it was found that any paper of American origin had the same effect, while paper of European or Japanese manufacture had no effect on the bugs!

Further investigation revealed that the juvenile hormone activity originated in the Balsam fir tree, the principal source of pulp for paper in Canada and the northern United States. The tree synthesizes a chemical (Fig. 5–14D) which accompanies the pulp all the way to the printed page. It, too, is a methyl ester of a certain unsaturated fatty acid derivative, and is very closely related to the other juvenile hormones. But, and this is the point, its action is selective against only one family of insects, the Pyrrhocoridae, which includes some of the most

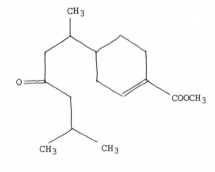

CH₂·CH₃

7
8   5
6   4
9   3
10   2   CH₃
O
11   1
CH₃   COOCH₃
CH₂
·
CH₃

(A) Natural juvenile hormone produced by male cecropia moth.
Carbon atoms, numbered from 1
to 11, are at each angle of
the structure, and are each
joined to 1 or 2 hydrogen
atoms (not shown). Letters
indicate atoms present at
terminals and branches.

CH₃

CH₃
O
CH₃
CH₃   CH₃   COOCH₃

(B) Synthetic analog of natural
cecropia hormone, made in
1965 by Department of Agriculture chemists. Note:
methyl (CH₃) groups have
replaced ethyl (CH₂·CH₃)
groups at top and lower
left of structure shown
in (A), but (B) is other-
wise identical to (A).

CH₃   Cl

CH₃
CH₃   CHOOCH₃
CH₃   CH₃
Cl

(C) One of the components of the
powerful synthetic hormone
mixture made from HCl and
farnesenic acid, and first
synthesized in Czechoslovakia.

CH₃
O
COOCH₃
CH₃   CH₃

(D) Natural hormone produced by
the Balsam fir tree, which has
strong juvenile hormone effect
on only one family of insects.

Fig. 5–14. Insect juvenile hormones. From Williams (25).

destructive pests of the cotton plant. Why the chemical is so selective is not known, but that it is suggests that juvenile hormones of other insects can be mimicked, and hence used as selective weapons against other pests.

Summary

In this chapter we have considered present methods of automotive transportation, production of electric power, and methods of insect control. In each case, we have explored less environmentally damaging alternatives, and indicated possibilities for the future. The obvious common thread has been our historical hurry, once a technology had been developed, to rush ahead and do the most economical thing on a grand scale, blissfully unaware that short-range convenience might turn into long-range disasters.

Yet each of these basic needs—transportation, electric power production, insect control—has in fact resulted in serious environmental damage, and could lead to future disasters if past practices should continue unchanged. The difference is that man has become aware of the effects he has had upon the environment, and is at least beginning to look for more ecologically attractive alternatives. These alternatives will emerge—have already, as exemplified by the work on insect sex attractants and juvenile hormones—if we are willing to expend the time, effort, and money to find them.

Literature Cited

1. John Macinko, "The Tailpipe Problem," *Environment Magazine*, Vol. 12, No. 5, June, 1970, pp. 6–13.
2. R.W. Hurn, "Mobile Combustion Sources," in *Air Pollution*, Arthur C. Stern, Ed., Vol. III, Academic Press, New York, 1968, Chap. 33.
3. Ruchlis, and Lemon, *Exploring Physics*, Harcourt, Brace and Co., New York, 1952 pp. 260–261.
4. National Report, "Your Car and Clean Air," Automobile Manufacturers Association, Detroit, Mich., pp. 7–8.
5. Allan T. Demaree, "Cars and Cities on a Collision Course," in *The Environment*, the Editors of *Fortune*, Perennial Library(P 189), Harper & Row, New York, 1970, pp. 89–90.
6. Andrew Jamison, *The Steam-Powered Automobile*, Indiana University Press, Bloomington, Ind., 1970, pp. 43–44.
7. Terri Aaronson, "Tempest over a Teapot," *Environment Magazine*, Vol. 11, No. 8, October, 1969, pp. 23–27.
8. George A. Hoffman, "The Electric Automobile," *Scientific American*, Vol. 215, No. 4, October, 1966, pp. 34–40.
9. Kurt Hohenemser, and Julian McCaull, "The Wind-Up Car," *Environment Magazine*, Vol. 12, No. 5, June, 1970, pp. 14–21, 32.
10. Robert W. Egbert, "Letter to the Editor," *Bulletin of the Atomic Scientists*, Vol. 26, No. 5, May, 1970, p. 43.
11. Sam H. Schurr, "Energy," *Scientific American*, Vol. 209, No. 3, Sept. 1963, pp. 111–126.

12. Committee on Resources & Man, National Academy of Sciences, National Research Council, *Resources & Man*, W.H. Freeman & Co., San Francisco, Calif., 1970, pp. 204–205.
13. Dean E. Abrahamson, *Environmental Cost of Electric Power*, Scientists' Institute For Public Information, New York, 1970, pp. 18–19.
14. Subcommittee on Environmental Improvement, *Cleaning Our Environment—The Chemical Basis for Action*, American Chemical Society, Washington, D.C., 1969, pp. 25, 64–74.
15. Glenn T. Seaborg, and Justin L. Bloom, "Fast Breeder Reactors," *Scientific American*, Vol. 223, No. 5, November, 1970, pp. 13–21.
16. Joel A. Snow, "Radioactivity from 'Clean' Nuclear Power," *Environment Magazine* (formerly *Scientist & Citizen*), Vol. 10, No. 4, May, 1968, pp. 97–101.
17. Richard H. Gilluly, "The Earth's Heat: A New Power Source," *Science News*, Vol. 98, No. 22, November 28, 1970, pp. 415–416.
18. Committee for Environmental Information, "The Price of Convenience," *Environment Magazine*, Vol. 12, No. 8, October, 1970, pp. 2–9.
19. Julian McCaull, and Mark Antell, "The Air of Safety," *Environment Magazine*, Vol. 12, No. 9, November, 1970, pp. 6–15.
20. Martin Jacobson, and Morton Beroza, "Insect Attractants," *Scientific American*, Vol. 211, No. 2, August, 1964, pp. 20–27.
21. Robert Van den Bosch, "Integrated Pest Control in California," *Bulletin of the Atomic Scientists*, Vol. 21, No. 3, March, 1965, pp. 22–26.
22. Jon Nordheimer, "Environmentalists Fight U.S. Spraying Plan on Fire Ants in South," *The New York Times*, December 13, 1970, p. 78.
23. George W. Irving, "Agricultural Pest Control and The Environment," *Science*, Vol. 168, No. 3938, June 19, 1970, pp. 1419–1424.
24. Martin Jacobson, and Morton Beroza, "Trapping Insects by Their Scents," *Chemistry*, Vol. 38, No. 6, June, 1965, pp. 7–11.
25. Carroll M. Williams, "Third-Generation Pesticides," *Scientific American*, Vol. 217, No. 1, July, 1967, pp. 13–17.

6

Summary:
The Cost of Environmental Improvement

Introduction

We began this study by tracing the development and growth of the pollution problem, and saw that it was a direct consequence of man's mushrooming technology and population. Obviously, this pollution was not deliberately planned, but resulted from our ignorance, oversights, and lack of basic understanding about the interdependence of all life on Earth. Indeed, man, after millions of years of existence here, has only in the past decade become aware of the magnitude and seriousness of the pollution problem he now faces. He is beginning to realize how his mistaken notion of an infinite Earth, with infinite resources and an infinite ability to cleanse itself, has led him to practices which—if continued—can only end in disaster for all life on this very finite world. A 1970 issue of *Life* magazine (1) warned:

Unless something is done to reverse environmental deterioration, say many qualified experts, horrors lie in wait. Others disagree, but scientists have solid experimental and theoretical evidence to support each of the following predictions:
- In a decade, urban dwellers will have to wear gas masks to survive air pollution.
- In the early 1980s, air pollution combined with a temperature inversion will kill thousands in some city of the United States.
- By 1985, air pollution will have reduced the amount of sunlight reaching Earth by one half.
- In the 1980s, a major ecological system—soil or water—will break down somewhere in the United States. New diseases that humans cannot resist will reach plague proportions.
- Increased carbon dioxide in the atmosphere will affect the Earth's temperature, leading to mass flooding or a new ice age.
- Rising noise levels will cause heart disease and hearing loss. Sonic booms from SSTs will damage children before birth.
- Residual DDT collecting in the human liver will make the use of certain common drugs dangerous and increase liver cancer.

Dire predictions, yet the second one has already come to pass, though perhaps on a smaller scale, in air pollution disasters already recorded. According to Italian scientists, the death of the Mediterranean is imminent, and the fourth of the *Life* predictions is about to be realized for some dozen cities along the east and west coasts of Italy (2). The key part in the quote above is, of course, the first sentence, for our assumption is that things *must* and *will* be done to reverse the present environmental deterioration. Indeed, we have attempted to suggest, in the preceding chapters, various technical solutions, or at least directions, which may help bring about this reversal. Most scientists believe that technical solutions to most pollution problems either are already available, or will be found through further study. There is less agreement, unfortunately, about whether man has the wisdom and will to re-order his priorities and habits so he can use his science and technology to deal effectively with these problems.

Before we look at specific examples of this apparent paradox, an analogy may help us to understand its existence. The analogy, which we may call "the tragedy of the commons," was first sketched by Lloyd well over 100 years ago (3), and reintroduced in this decade by biologist Garrett Hardin (4). A *commons* is simply a pasture open to all herdsmen, each of whom will try to keep as many cattle as possible on the commons. This "as many as possible" pattern has characterized man's behavior throughout recorded history, and it works fairly well as long as tribal wars and disease keep the numbers of both man and beast well below the carrying capacity of the land. What happens, however, when our technological progress conquers disease and allows almost unlimited expansion of the size of each herd of cattle? Each herdsman will realize it is to his advantage to add one more animal to his herd because he alone will receive all the proceeds from the sale of this additional animal, while the overgrazing of this animal is an effect shared by all herdsmen using the commons. Because he receives all the profits, while sharing only a fraction of the costs, each herdsman will add another animal to his herd, and another, and another. Thus each man is locked into a system that compels him to increase his herd without limit—in a world that is limited, so that the ability of the commons to support any cattle is soon exhausted (4).

Naive though it may be, this analogy applies not only to the ways in which men have denuded the Earth of its resources—be they grazing land, forests, minerals, or open space—but also to the ways in which men have polluted their environment. True, in this latter case, it is not a question of taking something out of the commons, but of putting something in—sewage, or chemical, radioactive, and heat wastes into water; noxious and dangerous fumes into the air; and litter and trash onto the face of the land. The reason the logic is the same is that, to date, rational man has found it cheaper to discharge his wastes into the commons than to purify them before release. Hence, the free-enterprise system, under which we have lived for so long, and which allows each man to pursue his own best interests, must be modified or changed to avoid disaster for all.

The reason some of our most knowledgeable scientists feel that science and technology alone will not solve the pollution problem is apparent from the analogy of the commons: Without a basic change in human values—the value of unlimited freedom for each individual, for example—we shall not have the will to use our scientific knowledge in ways that will benefit all men. It is worth noting that the pollution problem we now face is a direct consequence of overpopulation. Obviously, a large commons could easily support a small number of men and animals, no matter what they did. It made no difference how a lonely American frontiersman disposed of his waste for the simple reason that the myth of flowing water purifying itself every 10 miles was a reality as long as there were few people using the water. As population mushroomed, however, the natural chemical and biological recycling processes we have talked of became overloaded and broke down.

Thus, we not only face the prospect of ever-increasing controls over the freedom of individuals to use our common environment, but we must now face the fact that no peoples can be free to have children without limit. This limitation is unacceptable to many religions, and hence without basic changes in the areas of human values and morality, environmentally sound choices may never be made. The assumption here is that these changes *will* come because they *must*, and that our job is to help bring them about.

We shall therefore end our present study of chemical pollution by considering some specific examples of the kinds of choices we will be called upon to make, both now and in the years ahead, where the chemistry involved may be utilized to help solve the problem, but where the choice to use it must be dictated by new value judgments.

Nuclear Power Dilemma: Some Hard Choices

Increasingly in the days and years ahead, newspaper headlines will summarize legal battles between electric utilities and local government or conservation groups. Indeed, the action has already begun in the Chicago area, where two lawsuits have been brought against the Commonwealth Edison Company—one by the Metropolitan Sanitary District of Greater Chicago, another by 14 people, including a United Automobile Workers official (5). Both suits are seeking injunctions to restrain the utility from discharging heated, effluent water and radioactive material from a 2,200,000-kilowatt, two-unit nuclear power plant at Zion, Illinois. Both cite the threat of radioactive pollutants in Lake Michigan, the principal source of domestic water in a region of 8 million people. Even while these suits are pending in the Circuit Court of Cook County, Illinois, Commonwealth Edison is building additional nuclear power plants to provide badly needed electric generating capacity. In May 1973, Consolidated Edison of New York, after numerous delays, began operating a huge, 783,000-kilowatt nuclear power plant at Indian Point, 30 miles north of New York City. Six other nuclear power plants are either in operation, under construction, or planned for an area within a 100-mile radius of Philadelphia (Fig. 6—1).

Environmentalists concede that nuclear power could eliminate the sulfur dioxide air pollution from high-sulfur coal and oil burning, but point out that it introduces other pollutants (thermal and radioactive) into the environment. Thus the nuclear versus conventional power dilemma raises choices that require weighing the relative dangers of different types of pollutants. More than this, it raises basic questions about our human values and, hence, our priorities.

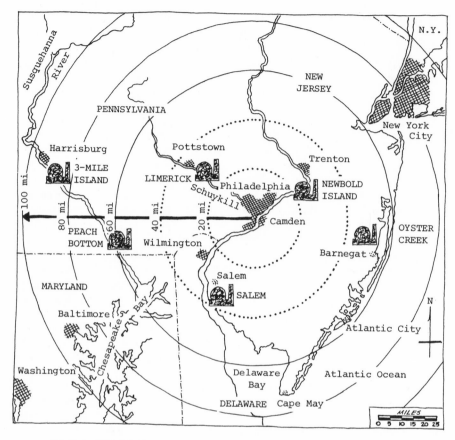

Fig. 6–1. Philadelphia area nuclear power plants in operation, under construction, or planned, as of February, 1971. From Ralph E. Lapp, "Nuclear Power Dilemma: With All Its Faults, We're Going to Need it." *The Philadelphia Inquirer*, February 14, 1971.

This is because America is at present, according to nuclear physicist Ralph Lapp (6), the most power-hungry nation in the world, and shows no signs of a letup in its ever-increasing demands for electricity. In 1900, for example, the annual per capita consumption of electricity per year amounted to 50 kilowatt-hours, enough energy to keep one 100-watt bulb burning for about 3 weeks. By the year 2000, Lapp estimates that per capita consumption will soar to 30,000 kilowatt-hours per year. This amounts to a 600-fold increase in energy consumption per capita, and while the population of the United States will have increased some four times by the year 2000, it is the 600-fold increase in individual energy consumption that is the crucial problem (6). Thus it is hardly surprising that power companies, especially on the East Coast, cannot get enough low-sulfur coal or fuel oil to meet new pollution control standards. And since natural gas is in short supply, utilities are turning to uranium as a new source of power.

It is therefore important for us to be knowledgeable about the risks of coal versus nuclear power generation; fears such as the possibility that a nuclear power plant might explode like an atom bomb are simply groundless because of the basic differences between reactors and atomic bombs (6). Other fears, however, are not so easily disposed of, and should be seriously considered. Nuclear engineers agree that the most probable form of nuclear power plant disaster would be a sudden loss of coolant liquid (used to cool the nuclear reactor core) because of defective materials or an accident. Not only would this radioactive coolant be a hazard if released to the environment (not likely, with proper precautions), but the uranium pellets might heat up enough to cause them to melt. A melt-down of this type occurred once in a nuclear plant near Monroe, Michigan, not far from Detroit. This was the Fermi reactor, using liquid sodium as a coolant. In 1966, accidental obstruction of the coolant caused an overheating of the fuel elements and a melt-down of some fuel.

Besides the thermal pollution problem, there is also the problem of radioactivity released to the environment, both during processing of reactor fuel and reactor operation. Finally, the problem of ultimate disposal of the spent but highly radioactive fuel elements is a very real one. Faced with the dangerous seismic activity surrounding their present burial site near Hanford, Washington, the Atomic Energy Commission (A.E.C.) has tentatively picked an area near Lyons, Kansas, where spent fuel could be buried in deep salt mines, the most tightly enclosed geologic formations known to man. A 1971 Kansas State geological report, however, was sharply critical of these A.E.C. plans, claiming that the possible effects of the radiation and heat upon the salt had not been sufficiently studied (7).

The Kansas study, based on both deep and shallow drillings at the proposed 1,000-acre site, concludes that while geologic conditions seem satisfactory, the problems relating to heat flow and surface subsidence remain largely unsolved. The A.E.C. solutions to these problems, the report says, have been based on simplified theoretical conditions, rather than on the actual geology of the mine area (8). Studies of radiation effects on salt show high heat-storage capabilities; nevertheless while the A.E.C. claims the salt would close about the radioactive spent-fuel containers and seal them in place, the Kansas geologists remain unconvinced. They feel that when the metal containers deteriorate (in about 6 months), followed by breakdown of the ceramic containers of the fuel several years later, the radioactive fuel could migrate through the salt and cause damage. They also worry about the interaction of subsidence, thermal expansion, and heat flow, which might break the seal of overlying rocks, permitting entry of ground waters that form the primary water source for the area (8).

Beyond these technical problems, and the risks of fossil fuel versus nuclear power, is the more fundamental problem. Since pollution in some form as an unpreventable consequence of power production apparently cannot be prevented, we must also make a choice between unlimited power production—which will bring with it increasingly unacceptable environmental problems—and using less power in efforts to preserve the environment as a whole. It is obvious that such a choice cannot be made from ignorance. It is also obvious that the present and future size of the world's human population is intimately related to energy needs. Although our United States population is a small fraction (5 or 6 percent) of the world's, we consume 40 percent of the world's energy supply. United Nations statistics from the early 1960s indicate that each of us consumed some 20 times the amount of energy as a person living in Latin America, Middle East, Africa, or Asia (9); this means that any reductions in energy use we can make will have corresponding effects on the entire global picture.

The relatively recently realized concept of a finite Earth, with finite energy resources, makes one wonder about the wisdom of the United States and Soviet space programs, for example, since the availability of enough energy for any significant amount of space activity is simply not in sight (10). And, as the under-developed countries begin to emerge and consume an ever-increasing share of the world's energy supplies, the problem will become more critical.

In their approach to energy problems, the interdisciplinary Study of Critical Environmental Problems (S.C.E.P.), sponsored in mid-1970 by the Massachusetts Institute of Technology, examined existing projections for responses to four principal questions:

1. What is the likely range of energy consumption in the United States and the world at present, in 1980, and in the year 2000?
2. What is the volume of pollutants likely to be generated, given various assumptions about how the necessary energy will be produced?
3. How is the production of these pollutants likely to be distributed around the globe?
4. What is the potential effectiveness of known pollution-control technology to reduce the pollutants' effects, and what order of magnitude difference could full use of these techniques have?

While these are certainly the kinds of questions that need to be answered if we are to make sensible choices, the S.C.E.P. group found that the present state of knowledge allowed a detailed response only to the first question, and even that only speculatively (11). Thus the continuing need for careful study, but the clear conclusion that we can no longer continue to accept unlimited side effects (i.e., pollution) of power production.

The SST Debate: The Priorities Used to Determine Today's Choices

The beginnings of the SuperSonic Transport (SST) project go back to late 1959, when an SST study group was formed within the Federal Aviation Administration (F.A.A.) to see if the new technology growing out of the proposed B-70 supersonic bomber project might be used to develop a commercial SST (12). While the B-70 project was abandoned as a bad idea in 1962, the SST project soon became a major United States government project in its own right. A joint French-British commitment to develop their own SST was a sharp competitive spur to Washington, and contributed to the U.S. decision in 1963 to proceed with building an SST that would be bigger, faster, and generally superior to any developed abroad, to be ready and flying by 1970.

First, there were design delays and, by 1967, considerable concern about the sonic boom such a faster-than-sound aircraft must produce. Resembling a clap of thunder, the boom is the result of the buildup of layers of air molecules which, since they can travel no faster than the speed of sound, simply cannot get out of the way of a supersonic plane. Thus an ever-increasing layer of squeezed-together air molecules form what amounts to a shock wave which, when it reaches the ground, produces the so-called sonic boom. Because of this phenomenon, which was recognized from the start by the SST project leaders, a prohibition against supersonic flight over land was considered likely, although such a ban would seriously restrict the SST's usefulness. Many knowledgeable observers, including the Federation of American Scientists (F.A.S.), feel that government rules on noise and sonic booms will be relaxed, if necessary, to keep finished SST aircraft economically viable. This was one of the factors that led to the F.A.S. policy statement opposing SST development (13).

In addition, there is widespread concern among scientists, as stated in the S.C.E.P. report (11), that flying the projected fleet of 500 SSTs will produce large enough amounts of exhaust gases and particulate matter (water vapor, carbon dioxide, and other gaseous and particulate pollutants) to affect, perhaps measurably, conditions in the stratosphere, as well as at lower altitudes. The S.C.E.P. report therefore recommended much additional further study, since they felt present knowledge did not permit accurate predictions of the extent of such effects. Besides the environmental concern expressed by many of the nation's most prestigious scientists, some 15 of the nation's top economists came out against SST development on purely economic grounds (14).

Such reservations were taken seriously enough by the Senate that they rejected funds for continued SST development by a 52 to 41 vote on December 3, 1970 (15). President Nixon bitterly denounced this action (16), stressing the waste of halting a half-completed project, and the necessity of the U.S. remaining first in the field of aviation. But the Administration's lobbying just before that Senate vote was even more revealing. Former test pilot William M. Magruder, heading the Administration's lobbying effort for the SST, blithely accused the 15 top economists who had opposed SST development of not understanding aviation economics, even though he himself is not an economist. Indeed, as was pointed out in an editorial of *The New York Times* (17), the airlines themselves are not anxious to buy SSTs, because an SST would cost twice as much as a 747 Jumbo jet, and would consume twice as much fuel while carrying only two-thirds as many passengers!

James J. Harford, executive secretary of the American Institute of Aeronautics, put forth some strange arguments just before that Senate vote. He asserted the SST would cause only 2.5 pounds of "overpressure," but the federal government itself had to stop sonic boom tests over Oklahoma City because overpressure of only 1.3 to 1.7 pounds provoked such painful noise levels. In an attempt to dismiss the ecological dangers, Mr. Harford pointed to a National Academy of Sciences study showing that 1,600 daily SST flights would produce about the same amount of water vapor—150,000 tons—as that produced by one large cloudburst in the tropics. And after all, he finished, "there are more than 3,000 cloudbursts around the globe each day!" (18). What Mr. Harford didn't say, if he understood it at all, was that cloudbursts do not put water vapor into the stratosphere, while the SST will! The SST, like all other jets, is a major polluter. In addition to the water vapor, a fleet of 500 SSTs will produce vast quantities of nitrogen oxides and other pollutants, and it is the utterly incalculable effects of these pollutants in the stratosphere that has aroused scientific concern.

Because of industry and Administration pressure, however, the House a week later refused to go along with the Senate vote (19), amid reports that the Nixon Administration actually suppressed criticism of the SST by other government agencies. According to one report (20), the Interior Department and Department of Health, Education and Welfare studies—required by law—were kept carefully out of the hands of Congress by the Administration during the debate that led to the crucial Senate vote. A few House members finally received the reports, but not until after the House declined to go along with the Senate rejection. Representative Paul Rogers, Democrat of Florida, charged that the Department of Transportation "deliberately delayed" releasing the documents until after the vote because of their potentially adverse effect (20). A similar charge was made several days later by Senators Nelson and Proxmire (21), after they had come into possession of a November 30th report, and Dec. 7th follow-up letter, from Health, Education and Welfare's Dr. Roger O. Egeberg, Assistant Secretary for Health and Scientific Affairs, to William Magruder, which Magruder apparently suppressed.

The tenor of the Administration's campaign to override the Senate—and proceed with SST development—shifted in early 1971 from environmental concerns to the issue of profit and aerospace industry unemployment (22). This, plus news that the Soviet Union had plans to put its supersonic TU-144 into commercial service between Moscow and Calcutta during the fall of 1971, reignited the sense of global competition. Such slogans as "Keep America FirSST" were developed by aerospace industry-labor coalitions, and culminated in full-page ads in major U.S. newspapers on March 8, 1971. The ad, sponsored by American Labor and Industry for the SST, was not only dishonest [with statements such as "Scientific studies indicate that our SSTs will not be harmful" (23)], but focused on the necessity of our staying ahead of the Soviets! Thus it is the tragedy of the commons all over again, but with one more SST, and another, and another! It is this kind of thinking that must change if we are to reverse the trend of environmental deterioration.

Aerosols and the Ozone Layer

Evidence of yet another type of technology-related pollution involving the stratospheric ozone layer has been accumulating since the mid-1970s. Man-made halomethane gases are collecting in the upper atmosphere in quantities that could significantly reduce the amount of stratospheric ozone. Since the ozone layer acts as a shield, absorbing some 99 percent of the deadly ultraviolet radiation from the Sun, the implications of these findings are potentially quite serious.

The two most common halomethanes are $CFCl_3$, the propellant in aerosol spray cans, and CF_2Cl_2, a refrigerant. Some 6×10^9 kg of these compounds has been produced prior to 1975—most of which has already been released into the atmosphere. Previously thought to be inert, these compounds are now known (since 1974) to dissociate into free Cl atoms by absorbing certain wavelengths of

The Cost of Environmental Improvement 109

ultraviolet light present only in the stratosphere. These free Cl atoms can catalytically destroy ozone (O_3) in the presence of ultraviolet (UV) light. A widely accepted mechanism would be:

$$Cl + O_3 \longrightarrow ClO + O_2$$
$$O_3 + UV \longrightarrow O_2 + O$$
$$\underline{ClO + O \longrightarrow Cl + O_2}$$

net reaction: $\quad 2O_3 + UV \longrightarrow 3O_2$

While it may be a long time before we will know positively the effects of our interference with the atmosphere, scientists are deeply concerned about the consequences of this ozone depletion.

Air and Waste-Water Treatment: Environmental Cleanup Costs

While many aspects of both water and air pollution control need further study, we already have the technology to remove many existing pollutants from our environment. The advanced water-treatment plant at Lake Tahoe, California, is typical of the facilities we could build today, providing we were willing to spend the money. Table 6–1 compares costs of the various treatment methods available today, and shows that advanced treatment methods will cost three to four times as much as the secondary treatment widely used in the U.S. today.

Besides being willing to accept the costs of effective water treatment, we must use our ever-increasing knowledge about the interdependence of living things on a finite Earth, and incorporate it into the design of such treatment procedures. The limited ecological foresight of oxidizing the phosphorus in sewage to phosphate ions, and then releasing them to the environment to hasten eutrophication of our lakes, shows how ineffective water treatment can be. First, we must distinguish between those wastes which contain a usable or potentially usable substance, and try to recover or re-use that substance, and wastes that are completely unwanted by-products of a purification process. Disposal of these latter wastes must be managed so as to minimize environmental damage. Uncontrolled disposal in deep wells, sometimes suggested as sound, may be disastrous because of the interconnectedness of the environment. Early attempts to dispose of citrus wastes underground created geysers of foul black liquid when anaerobic microbes produced methane and hydrogen sulfide from these wastes (23). More subtly, wastes injected at one site sometimes flow through underground strata to emerge elsewhere, and the possibility of such danger is suggested by the concept of interdependence.

Knowledge of the fate of pollutants in the environment could have prevented the widespread mercury pollution we now face, and can still be a powerful tool in minimizing the dangers of yet-to-be-disposed-of wastes. Disposal to the ocean takes advantage of the huge dilutions available, and is often the best method for such solubles as sodium and calcium chloride wastes, and magnesium sulfate brines, for they will be undetectable against the natural background of these same salts in the ocean. Their disposal on land, because their high solubility often washes them into fresh water supplies, is highly objectionable. But there are many substances which should never be discharged

Table 6-1. Comparing Cost and Quality of Various Water Treatment Methods*

| Treatment sequence | Estimated cumulative cost to treat 100 million gallons per day | | Permitted uses of treated water |
	Capital exp. (million $)	Operating exp. (¢/1,000 gal.)	
Raw waste	0	0	None; highly polluting.
Primary treatment	9.5	3.5	Partial pollution control; no direct reuse possible.
Secondary treatment (activated sludge)	20	8.3	Conventional pollution control; nonfood crop irrigation.
Coagulation-sedimentation	24	13	Improved pollution control; irrigation, recreation.
Carbon adsorption	30	17	Complete organic pollution control; swimming.
Electrodialysis	47	26	Complete inorganic-organic control.
Brine disposal	77	33	High-quality industrial supply; groundwater recharge.
Disinfection	77	34	Absolute pollution control; OK to drink.

*Condensed from a report by the Subcommittee on Environmental Improvement, "Cleaning Our Environment—The Chemical Basis for Action," American Chemical Society, Washington, D.C., 1969, pp. 124-125.

to the ocean, no matter how great the dilutions, because they are toxic and are concentrated as they pass up the food chains from plant to animal to predator and, ultimately, to man. Metals such as vanadium and cadmium, for example, are concentrated hundreds of thousands of times in this way; scallops can concentrate cadmium over two million times (23). Other painfully apparent examples are mercury and chlorinated hydrocarbon-type insecticides, such as DDT.

Combined sewage treatment is advantageous in some cases. Paper mill wastes and many soluble organic wastes are deficient in nitrogen, phosphates, or both. Municipal sewage, in contrast, normally contains more nitrogen and phosphorus than is needed for cell growth. To treat paper mill wastes by themselves would require purchase of nutrients for bacterial growth, and some of these nutrients would be released to the environment to hasten surface water eutrophication. But combined treatment of industrial and municipal wastes could be in everyone's best interests, provided industry pays its share of the larger treatment plant required (23).

One of the most serious air pollutants, sulfur dioxide, is emitted by stationary sources. Of the 25 or 26 million tons emitted into the atmosphere each year in the United States, over 80 percent is produced by the electric power utilities and industry, industry's contribution being mostly from power generation and smelting (24). We have seen that the sources of this SO_2 are the sulfur impurities present in coal and oil burned as fuel, which are oxidized to SO_2 during combustion. Most observers believe that large quantities of fossil fuels will continue to be burned to furnish electric power in the years ahead, so that sulfur removal at some point becomes ever more important.

Sulfur in coal occurs in sulfate, organic, and pyritic (FeS_2) forms, of which only sulfate sulfur is negligible. Coarse pyrite is readily removed by conventional cleaning processes, although these are less effective if the pyrite is finely disseminated. At present, some two-thirds of the coal mined in the United States is cleaned, most of it by specific gravity methods or flotation processes, both taking advantage of the fact that impurities are much more dense than the coal itself. Of course cleaning costs money, and does not remove all the sulfur, especially organic sulfur. Replacing coal with oil doesn't help, since desulfurizing the oil would almost double the $120 million capital investment of a typical oil refinery (24) if Caribbean residual fuel oil had its sulfur content reduced from 2.6 to 1 percent.

Since it is virtually impossible to remove all sulfur from fossil fuels, we should also look at the methods of removing SO_2 from the flue gases produced during burning itself. Four methods are considered economically feasible in the U.S. at present. The "Reinluft" process uses activated carbon to remove SO_2, after it has been oxidized to SO_3. This adsorbed SO_3 is later removed from the carbon by heating, and is recovered as sulfuric acid, which can then be sold to the chemical industry. The "alkalized alumina" process uses a mixture of 60 percent aluminum oxide and 40 percent sodium oxide to react with the sulfur dioxide, forming aluminum and sodium sulfates. The sulfates are then reduced by a combination of hydrogen and CO gas, to hydrogen sulfide (H_2S), which is finally converted to elemental sulfur. A newly developed process, known as "catalytic oxidation," is essentially a contact process sulfuric acid plant.

The well-known contact process uses platinum or vanadium pentoxide to catalyze the reaction

$$2SO_2(g) + O_2(g) \rightleftharpoons 2SO_3(g)$$

Table 6-2. Electric Power and Pollution Control Costs*

Service	Cost (mills/kilowatt-hour)
Electric Power	4.42
Particulate Control	0.06
Sulfur Dioxide removal	
Reinluft process	0.60
Alkalized alumina process	0.73
Catalytic oxidation process	0.64
Dolomite injection process	0.30

*From F.F. Aplan and Alfred J. Engel, "Sulfur Oxides and Power Generation," *Technology Tutor*, Vol. 1, No. 3, January, 1971, p. 73 (24).

The SO$_3$ produced is then converted to sulfuric acid and sold, as in the Reinluft process. Perhaps the simplest method of SO$_2$ removal is injection of the dry solid calcium magnesium carbonate, or dolomite, directly into a furnace. In the "dolomite injection" method, the dolomite is injected directly above the burning zone of a furnace, where it is decomposed to calcium and magnesium oxide. This material then catalyzes the oxidation of sulfur dioxide to sulfate ions, forming calcium and magnesium sulfate. These sulfates are useless waste products, and pose some ultimate disposal problems, but because of its simplicity, the dolomite injection method seems the most economical, and hence the most popular. Table 6–2 gives comparative costs, and shows that they will add about 15 percent to the consumer's electricity bill (24).

Summary

Our survey of the chemistry of pollution reveals that many aspects of pollution can be attacked with present technology, but this attack will take a great deal of money and the values shift that will allow us to spend it. According to the Sierra Club (25), the Organization for Economic Cooperation and Development made these rough cost estimates for cleaning and repairing the environment: A yearly appropriation of 2 percent of a nation's Gross National Product would only be enough to slow down deterioration; 4 percent might hold the line; and it would take up to 16 percent to achieve a real environmental cleanup. For the United States this would mean $150 billion a year. We now spend about $2.5 billion on environmental programs (25).

Yet if we are not willing to make this kind of commitment, we have already been given previews of what the future will hold. A simple example is what happened in New York City in the summer of 1970, as reported in *The New York Times* (26) (see Fig. 6–2). And the commitment is needed in all areas of the environment, as was poignantly revealed in this newspaper headline: "CITY HAS TRASH-DISPOSAL PLAN BUT STATE CAN'T FINANCE IT" (26). A glimpse of the consequences of no commitment was revealed by a joint National League of Cities—U.S. Conference of Mayors study, reported in *The New York Times* on June 10, 1973. According to that study, our cities are smothering in garbage, and almost half of them will run out of places to dump their trash within 5 years.

We have discussed the kinds of commitments and value shifts necessary to get started in earnest on environmental cleanup. Table 6–3 lists specific steps each individual can take. But the time for advice may be past and the time for action at hand. The warnings and advice have certainly been offered long enough to be carefully heeded. Just how long is revealed by an excerpt from the writings of the Hebrew philosopher and religious leader, Maimonides, who died in the year 1204. In his "Preservation of Youth" (27), Maimonides wrote:

The quality of urban air compared to desert or forest air is like turgid, polluted water, compared to pure, filtered water. This is because the pollution of the cities, caused by residents, their waste . . . fouls the air and creates . . . noxious gases. It is best, therefore, to live in a city with an open horizon . . . near forestation and outside water.

POLLUTED NEW YORK

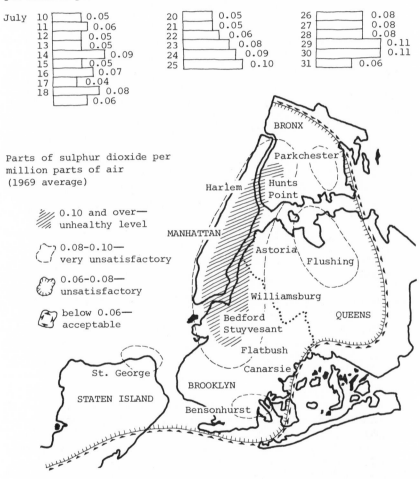

Citywide Average

Parts of sulphur dioxide
per million parts of air

July 10 — 0.05
11 — 0.06
12 — 0.05
13 — 0.05
14 — 0.09
15 — 0.05
16 — 0.07
17 — 0.04
18 — 0.08
— 0.06

20 — 0.05
21 — 0.05
22 — 0.06
23 — 0.08
24 — 0.09
25 — 0.10

26 — 0.08
27 — 0.08
28 — 0.08
29 — 0.11
30 — 0.11
31 — 0.06

Parts of sulphur dioxide per
million parts of air
(1969 average)

0.10 and over—
unhealthy level

0.08-0.10—
very unsatisfactory

0.06-0.08—
unsatisfactory

below 0.06—
acceptable

BRONX

Parkchester

Harlem   Hunts
Point

MANHATTAN

Astoria

Flushing

Williamsburg

Bedford
Stuyvesant   QUEENS

Flatbush

St. George   Canarsie

BROOKLYN

STATEN ISLAND

Bensonhurst

Fig. 6–2. Polluted New York, From *The New York Times,* Sunday, August 2,
1970, Section 4, p. 1.

Table 6–3. What You Can Do To Improve Your Environment*

The new law of interdependence has never been better focused. The carbon dioxide, the pesticides, the oil, the radiation threat—all infiltrate soil, air, wind, flora, fauna, and *food.* Where shall we start then if not in our own community? Here are some suggestions:

Water Pollution

1. Don't flush every time.
2. Don't put heavy paper, clothes, rags, disposable diapers, grease and solvents, or other chemicals into water disposal systems.
3. Use white toilet tissue—dyes pollute.
4. Don't fertilize your lawn—*runoff following rains pollute our streams and water table.*
5. Wash dishes and/or run your dishwasher once a day.
6. Object to TV soap commercials pushing high-phosphate content products. (For complete list, see *The New York Times,* 12/15/69.)
7. Use detergents low on phosphates which do not contain enzymes or use a biodegradable detergent. (Phosphates help algae and weeds grow and thus reduce oxygen level.) The six brands lowest in phosphate are:

Trend—1.4%	Cold Water All—9.8%
Diaper Pure—5.0%	Cold Power—19.9%
Wisk—7.6%	Fab—21.6%

8. Do not use full amounts called for on detergent boxes in dishwater, clothes washer, and/or dishwasher.
9. Promote regional sewage disposal systems; eliminate cesspools.
10. Enlist community groups to haul junk out of rivers, lakes, ponds, or streams.
11. Exert pressure on high-echelon factory officials to clean up any pollution-causing effluent or smoke stack emissions.
12. Write your state/local officials about evidence of pollution in your area.
13. Support your local conservation commission, open-space commission, and conservation organizations.
14. Most local colleges, county agents, can do tests of water for coliform counts, nitrates, etc.
15. Pressure your town and state officials to write and then enforce strict flood plain ordinances.
16. Write to federal officials: the President, Secretary of the Interior, your Senator and Congressman, on the following:
    a. Enforcement of current water pollution laws. Passage of new, stronger laws and stricter fines.
    b. Prohibition of dumping raw sewage, oil, herbicides, and garbage in all bodies of water, particularly the ocean.
    c. Prohibition of oil drilling along the Atlantic coast.
    d. State your opposition to the size of the super oil tankers.

*From Conservation and Environmental Science Center, Box 2230, R.D. 2, Browns Mills, N.J. 08015.

Table 6-3. What You Can Do To Improve Your Environment (Cont'd.)*

Air Pollution: A human needs 30 pounds of clean air a day. Doesn't his birthright entitle him to that?

1. Walk or bike whenever possible.
2. Avoid buying high-lead content gasoline or cars with engines requiring special high-octane mixtures. (Motor vehicles produce at least 60% of all air pollution.)
3. Make compost heaps (individual or community). Don't burn leaves.
4. Use and promote public transportation, car pools. Encourage the renaissance of commuter railroad lines where possible.
5. Voice your support for high-speed ground mass transit systems.
6. Monitor smoke stack emissions in your area. Contact local or state agencies for action.
7. Use less electricity, turn down your thermostat, and turn off unnecessary lights.
8. Do you really need that electric knife, can opener, electric broom, etc.?
9. Hang out your laundry in good weather and use lines in basements during bad weather.
10. Get a hand lawn mower—if there is a teenager and if you have a small lawn.
11. If you need a boat, get a canoe or sailboat.
12. A four- or six-cylinder car causes less air pollution.

Resources, Garbage, and Waste

1. Get a Boy Scout troop to collect newspapers, magazines, aluminum cans. (Re aluminum, write: Reynolds Metals Co., P.O. Box 2346-LI, Richmond, Virginia 23218.)
2. Encourage city officials to contract for separation of garbage.
3. Add garbage to your mulch heap. (Coffee grounds make excellent mulch.)
4. Don't use plastics of polyvinyl chloride, such as Saran. Complain to store manager if such items as bananas, cucumbers, and peppers are unnecessarily plastic-wrapped. Buy unwrapped produce. (OK—polystyrene, polyethylene, foam-types.)
5. Use wax paper and cellophane.
6. Don't use aluminum wrap or cans unless your community collects them.
7. Take a shopping bag to stores. You pay twice for all multisize bags.
8. Buy products which conserve on wrappings—you pay for the wrappings.
9. Go easy on paper towels, paper cups, and paper napkins.
10. Junk mail makes good drawing paper for children.
11. Save—Christmas wrappings, string, etc.
12. We throw out 48 billion cans per year, and the number is increasing. (Shouldn't we push for taxes on beer and soft drinks to compensate for the cost of cleaning up highway litter? It is easier in the long run.)
13. *Stifle* that inner voice that says this little bit won't hurt. (Multiply it by 203,000,000.)

*From Conservation and Environmental Science Center, Box 2230, R.D. 2, Browns Mills, N.J. 08015.

Table 6—3. What You Can Do To Improve Your Environment (Cont'd.)*

Pesticides

Who has decided—who has the right to decide—for the countless legions of people who were not consulted that the supreme value is a world without insects, even though it be also a sterile world ungraced by the curving wing of a bird in flight?

Rachel Carson, *Silent Spring*, 1962

1. Do not use long-lived pesticides, chlorinated hydrocarbons, such as DDT, Dieldrin, Aldrin, Endrin, Heptachlor, Chlordane, Lindane.
2. Do use short-lived pesticides, such as Malathion, Rotenone, Off, Cutters Lotion.
3. Dispose of all your existing hard pesticides by burning them at an incinerator reaching a temperature of at least $1,800°$F. The incinerator should be equipped with water scrubbers. (Small quantities may be buried in a remote dry area with a low water table downgrade from houses, ponds, water supplies, etc.) Combine the pesticide equally with calcium hydroxide (slaked lime) and then empty it into a pit. Cover the top of the mixture with the lime. The top of the layer should be at least 18 inches below the ground surface level. Mound the top so that water will run off the sides.
4. Do not use fertilizer with arsenate (arsenic).
5. Plant trees and shrubs to attract birds.
6. Voice your objection to the Board of Health about mass spraying by mosquito control commissions, particularly near areas such as lakes, ponds, marshes, etc.
7. Introduce lady bugs, gambusia (mosquito-eating fish), praying mantis, and dragon flies in such areas.
8. Build purple martin houses. Each purple martin can eat 1,000 mosquitos a day.

*From Conservation and Environmental Science Center, Box 2230, R.D. 2, Browns Mills, N.J. 08015.

Literature Cited

1. The Editors, "Ecology," *Life*, Vol. 68, No. 3, January 30, 1970.
2. John Cornwell, "Is the Mediterranean Dying?", *The New York Times Magazine*, February 21, 1971.
3. W.F. Lloyd, *Two Lectures on the Checks to Population*, Oxford University Press, Oxford, England, 1833.
4. Garrett Hardin, "The Tragedy of the Commons," *Science*, Vol. 162, December 13, 1968, pp. 1243—1248.
5. Philip F. Gustafson, "Nuclear Power and Thermal Pollution: Zion, Illinois," *Bulletin of the Atomic Scientists*, Vol. 26, No. 3, March, 1970, pp. 17—23.
6. Ralph E. Lapp, "The Four Big Fears About Nuclear Power," *The New York Times Magazine*, February 7, 1971.

7. Anthony Ripley, "Kansas Geologists Oppose a Nuclear Waste Dump," *The New York Times*, February 17, 1971, p. 27.
8. The Editors, "The Kansas Geologists and the AEC," *Science News*, Vol. 99, No. 10, March 6, 1971, p. 161.
9. Sam H. Schurr, "Energy," *Scientific American*, Vol. 209, No. 3, September, 1963, pp. 111–125.
10. Howard T. Odum, *Environment, Power, and Society*, John Wiley & Sons, New York, 1971, p. 304.
11. Report of the Study of Critical Environmental Problems (S.C.E.P.), "Man's Impact on the Global Environment," M.I.T. Press, Cambridge, Mass., 1970.
12. News and Comment, "SST: Commercial Race or Technology Experiment?", *Science*, Vol. 169, No. 3943, July 24, 1970, pp. 352–355.
13. Policy Statement: "Federation Opposes the SST," *F.A.S. Newsletter*, Vol. 23, No. 8, November, 1970, p. 3.
14. News and Comment, "15 Top Economists Oppose SST," *Science*, Vol. 169, No. 3952, September 25, 1970, p. 1292.
15. Christopher Lydon, "Senate Rejects SST Fund by 52-41 Vote After Drive by Environmental Lobby," *The New York Times*, December 4, 1970, p. 1.
16. John W. Finney, "President Bids Congress Reverse Rejection of SST," *The New York Times*, December 6, 1970, p. 1.
17. Editorial, "Downwind From The SST," *The New York Times*, December 2, 1970.
18. James J. Harford, "Up The SST," *The New York Times*, December 1, 1970.
19. Christopher Lydon, "House Declines to Back Senate in SST Fund Curb," *The New York Times*, December 9, 1970.
20. Clarke Hoyt, "Agencies' Criticism of SST Kept from Congress," *Philadelphia Inquirer*, December 9, 1970.
21. E.W. Kenworthy, "2 SST Opponents Charge Administration Supressed Criticism," *The New York Times*, December 11, 1970.
22. Christopher Lydon, "Backers of Supersonic Plane Making Strong Comeback as Issue Shifts from Environment to Profits," *The New York Times*, March 1, 1971.
23. Robert B. Dean, "Ultimate Disposal of Industrial Waste: An Overview," *Technology Review*, Vol. 73, No. 5, March, 1971, pp. 20–25.
24. F.F. Aplan, and Alfred J. Engel, "Sulfur Oxides and Power Generation," *Technology Tutor*, Vol. 1, No. 3, January, 1971, pp. 67–73.
25. The Editors, "Environmental Repairs," *Sierra Club News*, April, 1970, p. 22.
26. Anthony Lane, "City Has Trash-Disposal Plan But State Can't Finance It," *Philadelphia Inquirer*, March 2, 1971.
27. Gilbert S. Rosenthal, Ed., *Maimonides: His Wisdom for our Time*, Funk and Wagnalls (A Sabra Book), New York, 1969, p. 13.

Index